Air Force Nonrated Technical Training

Selected Topics to Improve Efficiency

KATHLEEN REEDY, LISA M. HARRINGTON, BART E. BENNETT,
BARBARA BICKSLER, JAMES R. BROYLES, ROBERT CORSI,
PAUL EMSLIE, CHARLES A. GOLDMAN, DANIEL IBARRA,
DARRELL D. JONES, RITA KARAM, MATTHEW WALSH

Prepared for the United States Air Force
Approved for public release; distribution unlimited

For more information on this publication, visit www.rand.org/t/RR2774

Library of Congress Cataloging-in-Publication Data is available for this publication.
ISBN: 978-1-9774-0498-5

Published by the RAND Corporation, Santa Monica, Calif.
© Copyright 2020 RAND Corporation
RAND® is a registered trademark.

Limited Print and Electronic Distribution Rights

This document and trademark(s) contained herein are protected by law. This representation of RAND intellectual property is provided for noncommercial use only. Unauthorized posting of this publication online is prohibited. Permission is given to duplicate this document for personal use only, as long as it is unaltered and complete. Permission is required from RAND to reproduce, or reuse in another form, any of its research documents for commercial use. For information on reprint and linking permissions, please visit www.rand.org/pubs/permissions.

The RAND Corporation is a research organization that develops solutions to public policy challenges to help make communities throughout the world safer and more secure, healthier and more prosperous. RAND is nonprofit, nonpartisan, and committed to the public interest.

RAND's publications do not necessarily reflect the opinions of its research clients and sponsors.

Support RAND
Make a tax-deductible charitable contribution at
www.rand.org/giving/contribute

www.rand.org

Preface

This report is part of continuing research that began in fiscal year (FY) 2016 by RAND Project AIR FORCE (PAF) for the Air Force Air Education and Training Command (AETC). In FY 2016, the RAND Corporation conducted a broad review of the nonrated technical training pipeline, with the goal of identifying pipeline inefficiencies that, if corrected, could lead to a more efficient and flexible enterprise. PAF identified opportunities to improve the planning process, resource-allocation process, and the flow of students through the technical training pipeline.

As a follow-on to this effort, AETC asked PAF to look outside the Air Force for insights and best practices upon which they could draw. AETC identified three particular topics of interest: (1) how colleges and universities right-size their instructor corps in the face of fluctuations in enrollments and demands for coursework, (2) best practices associated with supply chain management, and (3) approaches for developing a flexible instructor pool. Each of these topics is discussed in a separate part of this volume.

At the conclusion of our research, common threads emerged from what might appear to be rather disparate topics. One is the realization that there is no one-size-fits-all model that will work AETC-wide to achieve more-efficient operations. Instead, these concepts and the resource decisions that they drive are best applied to the individual training pipelines for each Air Force specialty. The second thread is that flexibility across the training pipeline is the key to improving planning and resource efficiency.

The research reported here was commissioned by the Vice Commander, AETC/CV, and conducted within the Manpower, Personnel, and Training Program of RAND Project AIR FORCE as part of a fiscal year 2017 project "Methods to Evaluate Changes to Training."

We welcome your questions and comments regarding this report, which can be addressed to the lead author as follows:

Lisa Harrington, RAND Corporation, 1200 South Hayes Street, Arlington, Virginia, 22202, 703-413-1100 x5255, lharring@rand.org.

RAND Project AIR FORCE

RAND Project AIR FORCE (PAF), a division of the RAND Corporation, is the U.S. Air Force's federally funded research and development center for studies and analyses. PAF provides the Air Force with independent analyses of policy alternatives affecting the development, employment, combat readiness, and support of current and future air, space, and cyber forces. Research is conducted in four programs: Strategy and Doctrine; Force

Modernization and Employment; Manpower, Personnel, and Training; and Resource Management. The research reported here was prepared under contract FA7014-16-D-1000.

Additional information about PAF is available on our website: www.rand.org/paf/

This report documents work originally shared with the U.S. Air Force on October 11, 2017. The draft report, issued on September 29, 2017, was reviewed by formal peer reviewers and U.S. Air Force subject-matter experts.

Contents

Preface iii
Figures viii
Tables ix
Summary x
Acknowledgments xv
Abbreviations xvi
Part I. Introduction 1
1. Improving Efficiency in the Air Force Nonrated Technical Training Pipeline 2
Part II. What Can the Air Force Learn from Higher Education Institutions About
    Meeting Challenges in Technical Training? 5
2. Introduction and Overview of American Community Colleges 6
    Overview of American Community Colleges 7
3. Outsourcing and Automating Services Ancillary to the Core Mission 9
    Outsourcing or Contracting: Software Improvements and Automation
       in Air Force Technical Training 10
4. Anticipating Changes in Workforce Demand 12
    Estimating Workforce Demand in Air Force Technical Training 13
5. Adjusting Capacity in Response to Changes in Demand 15
    Using Adjunct Faculty to Adjust Air Force Technical Training Capacity 16
6. Alternative Methods for Instructional Delivery 19
    Online and Blended Learning 19
    Competency-Based Education 20
    Effectiveness of Alternative Instruction Methods 21
    Application of Alternative Instruction Methods to Air Force Technical Training 23
    Final Considerations for Adopting Alternative Instruction Methods 26
7. Conclusion 28
Part III. A Supply Chain and Production Systems Approach
    to Programming and Resourcing 30
8. Introduction 31
9. Parallels Between the Nonrated Technical Training Process
    and Supply Chain and Production Systems 33
    Instructor Production System and Reactive Resourcing 34
    Trainee Production System and Flexible Resourcing 38
10. Resourcing Approaches to Supply Chain and Production Systems 40
    Robustness as an Enabler to Supply Chain Resilience 40
    Agility as an Enabler to Supply Chain Resilience 41

Deciding Between a Robust and an Agile Resourcing Strategy 42
11. A Proposed Process for Choosing Resourcing Strategies 44
    Step 1: Categorize AFSCs 44
    Step 2: Quantify Resourcing Bounds 45
        Notional Example of Step 2 46
        Options for Increasing Instructor Availability 47
    Step 3: Determine Potential for Requirement Changes 49
    Step 4: Estimate Effects of Potential Resourcing Strategies 52
    Step 5. Choose a Resourcing Strategy 53
    Example of the Proposed Resourcing Process: Intelligence Officer Course (AFSC 14N1) 54
12. Summary and Recommendations 58
Part IV. An Approach for Developing an Agile Technical Training Instructor Mix 60
13. The Case for Building a Flexible Instructor Approach 61
14. Instructor Pool Elements 64
    A. AETC Assigned Active Duty Instructors 64
    B. Retired Active Duty Instructors (Limited Recall) 64
    C. Prior-Service Instructors (Left Active Duty, Not Retired) 65
    D. Currently Employed Air Force Civilians 65
    E. Civilians with Instructor Experience (Retired Annuitants) 66
    F. Guard and Reserve (Active Guard and Reserve) 66
    G. Guard and Reserve (Traditional Reservists) 66
    H. Guard and Reserve (Limited Active Duty Recall) 66
    I. Prior Active-Duty Instructors (Unit Assigned) 67
    J. Contractors 67
    K. AETC Overmanning 67
    L. Adjunct Military and Civilian Instructors 68
    M. Nongovernment Adjunct Instructors 68
15. Instructor Suitability Evaluation 70
    Creating an Instructor Pool Using the Assessment Criteria 73
        A Notional Example 74
        Recommendations 75
Part V. Concluding Thoughts 76
16. The Benefits of Flexibility for Improving Technical Training Efficiency 77
17. Where Should the Air Force Go from Here? 84
    Employing the Concepts Reviewed 85
        Planning 85
        Instructor Mix 85
        Content Delivery 86
        Quality Control 86
    Conclusion 87
Appendix A. Statistics of Air Force Special Code Program Guidance Letter Changes 88

Appendix B. Officer Air Force Specialty Code Program Guidance Letter Changes 93
References 94

Figures

Figure 9.1. A Conceptual Representation of the Instructor and the Trainee Supply Chains 34
Figure 9.2. Current Instructor Make-to-Order Production System 35
Figure 9.3. Percentage of Enlisted Air Force Specialty Codes with Increases in the
    Trained Personnel Requirement as Stated in the Program Guidance Letter 36
Figure 9.4. Proposed Instructor Made-to-Stock Production System 37
Figure 9.5. Trainee Production System for Guaranteed Training Enlistment
    Program and Non–Guaranteed Training Enlistment Program Airmen 38
Figure 11.1. Notional Example of Binding Resource Capacity 46
Figure 11.2. Notional Example of Students in the Pipeline Given Accession
    or Requirement Changes and Different Resourcing Approaches 53
Figure 11.3. Example of Maximum Cumulative Number of Graduates from
    Fiscal Year 2017 Intelligence Officer (AFSC 14N1) Class, by Class
    End Date Given Course of Action Selected 57
Figure 16.1. Distribution of Class Sizes for Seating Options for the
    Maintenance Apprentice (F-16) Part 1 Course 81
Figure 16.2. Cumulative SAT Days for Seating Options for the Maintenance
    Apprentice (F-16) Part 1 Course 82
Figure B.1. Percentage of Air Force Specialty Codes with
    Program Guidance Letter Changes 93

Tables

Table 11.1. Number of AFSCs in Each PGL Change Category 50
Table 11.2. Requirements Changes for Cryptologic Language Analyst 51
Table 11.3. Notional Example of Requirement Change Scenarios 51
Table 11.4. Fiscal Year 2017 Intelligence Officer (AFSC 14N1) Classes,
    as of May 12, 2017 55
Table 15.1. Option Assessment Criteria 70
Table 15.2. Assessment of Instructor Option Suitability by Criteria 71
Table 16.1. Data for Maintenance Apprentice (F-16) Part 1 79
Table 16.2. Estimated Outcomes for the Idealized Case for the
    Maintenance Apprentice (F-16) Part 1 Course 79
Table 16.3. Number of Class Starts for Seating Options for the
    Maintenance Apprentice (F-16) Part 1 Course 81
Table A.1. Enlisted Air Force Specialty Codes in Category 1 with
    Program Guidance Letter Change Data and Statistics 89
Table A.2. Enlisted Air Force Specialty Codes in Categories 2, 3, or 4
    with Program Guidance Letter Change Data and Statistics 91
Table A.3. Officer Air Force Specialty Codes in All Categories with
    Program Guidance Letter Change Data and Statistics 92

Summary

In fiscal year (FY) 2016, RAND Project AIR FORCE conducted a study for the Air Force Air Education and Training Command (AETC) of inefficiencies in the nonrated technical training pipeline. The goal of this research was to identify ways to improve the efficiency and responsiveness of the Air Force's technical training enterprise for both officers and enlisted personnel. The study identified many opportunities to improve the technical training planning process, resource allocation process, and the flow of students through the technical training pipeline. Perhaps the most significant challenge that we identified was how difficult it is for the process overall to respond to changes in the demand for trained airmen, particularly when requirements change in the year of execution. The inability to rapidly right-size the instructor pool was a central shortfall.

With the results of this effort in hand, AETC began to look outside the Air Force for insights and best practices upon which they could draw. AETC asked us to investigate two models of particular interest: (1) how colleges and universities, institutions that similarly face fluctuations in enrollments and demands for coursework, right-size their instructor corps and (2) best practices associated with supply chain management. In addition, AETC asked us to explore one potential solution—a flexible instructor pool—to address the challenging problem of rapidly sizing the instructor corps to respond to changing requirements that occur late in the planning process.

Each of these three topics, as posed by the Air Force, is examined in a separate part of this report. In this summary, we offer a brief overview of the findings and recommendations based on these analyses and concluding thoughts on the benefits the Air Force can realize from implementing the types of flexible resourcing options discussed.

What Can the Air Force Learn from Higher-Education Institutions?

Colleges and universities face many of the same challenges as the Air Force technical training enterprise, and, therefore, some of the practices that colleges have adopted to meet these challenges might also hold promise for the Air Force. The two main advantages that colleges have over AETC in terms of planning for and providing education is their ability to rapidly flex when demands change and to more readily adopt new methods of planning and instruction. The use of adjunct instructors, combined with online and blended learning in particular, allows these institutions to quickly adjust the size of their workforce to respond to unexpected surges of students without having to make long-term commitments to additional staff. In addition, community colleges focused on technical degrees make a strong effort to work with the local community to identify educational needs and expected throughput in certain career fields.

Finally, colleges and universities can also take advantage of outside expertise by outsourcing parts of their registrar functions and can quickly adopt the latest technological advances to enhance the training and planning processes. Although there are differences between colleges and universities and the Air Force's training mission, AETC may benefit from adopting or adapting some of the practices currently being employed in postsecondary education in the following areas.

- Planning and resourcing approaches
  - *Outsource or contract for improved software for administrative functions*: The Air Force could consider leveraging private-sector expertise to either employ or adapt existing types of software to the Air Force environment to help reduce the complexities of scheduling and tracking airmen's progress.
  - *Improve planning with modeling and improved communication:* AETC could get ahead of some of the uncertainty in its planning system by coordinating with Air Force, Manpower, Personnel and Services (AF/A1); Second Air Force (2AF); functional managers; and the training squadrons to develop models that may help informally assess levels of future need and therefore give AETC more lead time to plan.
  - *Develop an adjunct-type model for the instructor corps:* This type of model will help provide additional capacity and support during periods of student surges without requiring AETC to bring on additional full-time instructors.
- Instructional approaches
  - *Explore the use of online or blended learning for initial skills training:* Incorporating online or blended learning into the curriculum offers flexibility to use instructors who may not be colocated with the schoolhouse or to have students take a portion of their training from other locations, such as their first duty station.
  - *Evaluate the use of competency-based learning:* Competency-based learning can be an effective educational approach that, when combined with online or hybrid learning, may decrease the instruction time students require before graduation.

A Supply Chain and Production-Systems Approach to Programming and Resourcing

In general, supply chains are characterized by suppliers providing products to consumer organizations, to which consumer organizations add value to their products, and then sell these products to their customers. The Air Force nonrated technical training process can be seen as two supply chains that have different underlying production systems: a supply chain for creating instructors and a supply chain for creating trained airmen. Supply chain management strives to integrate information and cooperation across the supply chain echelons to enable better planning and less uncertainty in materiel availability and customer demand.

The analogy between commercial supply chains and the "production" of trained airmen is not a perfect one. Many supply chain management principles have direct application to managing technical training, particularly those pertaining to resourcing strategies. Drawing on best practices found in the supply chain and production literature, we developed a resourcing decisionmaking framework specific to the technical training environment and proposed a research and analysis plan to support the framework. The framework uses a risk-management approach to stratify resourcing decisions based on the level of uncertainty by employing a five-step process: (1) categorize pipelines, (2) quantify resourcing bounds, (3) estimate probability for requirement changes, (4) estimate effects of potential resourcing strategies, and (5) choose a resourcing strategy.

Steps 1, 2, and 3 are information-gathering activities to identify which Air Force Specialty Codes (AFSCs) need to respond to potential requirement changes, the magnitude and timeliness of adding additional resources, and the possibility of requirement changes. Step 4 is an analysis and modeling step in which the information gathered is combined and hypothetical resourcing approaches are modeled to estimate effects on the number of graduates, students awaiting training, and other performance metrics. In Step 5, the resourcing-approach decision is made by taking into consideration the information obtained and the modeling results along with external information about cost and desirability to add additional resources. Although the Air Force has implemented some elements of this process, it has yet to take a comprehensive and systematic approach to resourcing decisionmaking that is informed by best practices.

Several related recommendations will directly support the resilient resourcing approach presented in this report:

- *Develop a capacity-visibility capability where estimates are accurate, updated regularly, and include feasible timelines and magnitude of available additional resources.* Capacity information should be readily accessible to all relevant organizations, including Headquarters Air Force, AETC, 2AF, training squadrons, and schoolhouses. (This capability will enable Step 2.)
- *Communicate potential requirement uncertainties in addition to a single planning target.* These uncertainties can be expressed as ranges around the Program Guidance Letter quantities and can be used to proactively assess whether additional resources should be allocated and on what timeline to meet potential requirement changes. (This concept is incorporated in Step 3.)
- *Develop technical training modeling capabilities.* Models can evaluate the pipeline effect of resourcing strategies given potential requirement changes; they can be used to manage leadership expectations on response timeliness and inform resourcing decision. (This is Step 4.)
- *Create a single office that is responsible for gathering all information*, modeling the effects given resource feasibility and potential requirement changes, and making the decision on the appropriate resourcing approach. A single office will centralize information collection and enable informed decisions about the resourcing approach.

Flexible Instructor Pool for AETC Technical Training

The Air Force training vision for the future calls for a complete review of how content is developed and delivered and a complete review of what category of instructors are best suited to deliver the training. Its aim is to develop a training system that can rapidly react to changes in the number of students required without compromising training quality, within the budget approved for initial skills training. A flexible instructor pool is a new approach to instructor resourcing that would enable AETC to quickly react to changing student requirements. Such a pool would draw on a mix of active duty, reserve component, civilians, contractors, and adjunct instructors—options that could be used alone or in combinations to expand instructor capacity, as needed to respond to changes in demand.

To determine the proper instructor pool, a review of possible instructor options is required. We identified and evaluated 15 options that range from active-duty instructors to full-time guard or reservists to more innovative concepts such as remote instruction. To evaluate each option, we considered various assessment criteria such as cost, flexibility, quality, stability, surge capability, planning lead time, short-term technical ability, unit impact, and end strength impact. Understanding how these different criteria apply to each instructor option and to the options relative to one another is essential to determine how best to meet AETC's instructor requirements. There may be circumstances, for example, where a relatively higher-cost option offers the most flexibility, and that cost may be worthwhile when timing is a priority or particular technical knowledge is difficult to tap into.

Not all the options we evaluated are available to the Air Force today. But our analysis illustrates the first steps toward developing a strategy for creating a pool of available instructors on which the schoolhouses can draw. Although we evaluate the flexible instructor options against a set of criteria, such an evaluation must be conducted within the context of other changes that AETC could make in delivering initial skills training to its airmen, such as implementing a blended learning concept or incorporating enabling technology at its educational institutions to support both instruction and administration.

We recommend that AETC use a systematic approach, such as the one demonstrated in the report, to investigate options for creating a pool of instructors when it is determined that a proactive resourcing strategy is appropriate or when immediate needs for instructors increase. Such an approach should

- Identify potential sources of instructors.
- Ensure assessment criteria represent the characteristics desired of such a pool.
- Systematically identify barriers to using particular categories of manpower for an instructor pool and develop policies to mitigate.
- Develop pilot programs for promising sources of instructor manpower for a small set of AFSCs with critical shortages. Ensure results are fed back into the process for identifying and assessing options.

Conclusions

The research reported here focused on options that the Air Force might consider to improve the adaptability of the technical training system. Although the concepts discussed differ, two themes emerged. First, there is no one-size-fits-all model that will work AETC-wide. Instead, these concepts, along with commensurate resourcing decisions, should be made for each training pipeline to optimize the use of resources and improve pipeline efficiency. Second, flexibility across the training pipeline is the key to improving planning and resourcing efficiency. Without the flexibility to run different parts of the technical training pipeline in different ways or to conduct regular analysis that leads to adjustments to the process over time, the inefficiencies created by inflexibility will continue to cost the Air Force time and money in very real ways.

Together, the topics discussed in this report offer options and concepts that the Air Force could adopt to improve flexibility and responsiveness in the technical training pipeline. Implementing these approaches will require investments in time and resources, but the benefits can be significant and help the Air Force reach its goal of developing a more flexible and responsive technical training enterprise.

Acknowledgments

We are grateful to many people who were involved in this research. In particular, we would like to thank our Air Force research sponsor, Maj Gen Mark Brown, Deputy Commander, Air Education and Training Command, and action officer Col Michael Grunwald for their strategic guidance and perspectives throughout this study, as well as Col Timothy Owens, Angela Canada, and Sherry Hernandez, who provided historical background and key Air Force viewpoints. We also thank Kent Miller at the Navy's Production Management Office for providing the naval perspective on initial skills training.

We are also grateful to the staff at the Second Air Force, who helped us understand their role in the pipeline process. We appreciate the information provided by the various leaders and representatives from the training wings, groups, and squadrons; Headquarters Air Force; the Air Force Personnel Center; and associates at Northern Virginia Community College. This research was enriched by their contributions.

This research also benefited from helpful insights and comments provided by several colleagues at the RAND Corporation, including Ray Conley and John Ausink. We also thank our reviewers, Sean Robson and Marc Robbins, for their thoughtful comments that greatly improved this report.

Abbreviations

14NX	Intelligence Officer (job title within the Air Force Specialty Code)
2AF	Second Air Force
ADSS	Air Education and Training Command Decision Support System
AETC	Air Education and Training Command
AF/A1	Air Force, Manpower, Personnel and Services
AFRC	Air Force Reserve Command
AFSC	Air Force Specialty Code
AGR	active guard and reserve
ATO	assemble to order
BMT	Basic Military Training
DoD	U.S. Department of Defense
DSD	Developmental Special Duty
FY	fiscal year
GTEP	Guaranteed Training Enlistment Program
IT	information technology
MTO	make to order
MTS	make to stock
PAF	RAND Project AIR FORCE
PGL	Program Guidance Letter
POM	Program Objective Memorandum
SAT	student awaiting training
TA	teaching assistant
TDY	temporary duty

Part I. Introduction

Lisa M. Harrington, Kathleen Reedy, and Barbara Bicksler

1. Improving Efficiency in the Air Force Nonrated Technical Training Pipeline

The quantity and quality of trained personnel directly affects combat capability, and the supply of sufficiently trained Air Force personnel relies on effective management of training production. Air Education and Training Command (AETC) must develop and maintain a responsive training pipeline in a constantly evolving environment. A wide range of factors influences the supply and demand for personnel, such as the need for a rapid buildup of forces, acquisition of technologies or systems that reduce manpower requirements or change the types of competencies required, restructuring of career fields, the privatization or outsourcing of skills, and various other force-management decisions. Planning, programming, and managing the training pipeline must accommodate changes in specialties, student throughput, curriculum, the costs of resources, and the presence of and access to an instructor pool, as well as classroom capacity, training equipment, student housing, and a wide range of base operating support.

During fiscal year (FY) 2016, AETC Directorate of Intelligence, Operations and Nuclear Integration (AETC A2/3/10) asked RAND Project AIR FORCE (PAF) to undertake a study of pipeline inefficiencies and ways to improve the efficiency and responsiveness of the Air Force's nonrated technical training processes—both officer and enlisted. The goal of the study was to identify opportunities for optimizing overall processes at all levels of technical training—corporate Air Force planning and programming; training management at AETC, Second Air Force (2AF), and individual training wings; and the provision of training in individual courses—and to recommend policy and process changes that could improve overall efficiency.

We identified several opportunities to improve efficiency and responsiveness within the Air Force's technical training enterprise—*efficiency* in terms of maximizing student throughput and *agility* in terms of responding to unforeseen changes in demand. Our analysis concluded that inefficiencies exist in the technical training planning process, in the resource-allocation process, and in the flow of students through the basic and technical training pipeline. Some of the major challenges AETC faces include effectively planning for student throughput, adjusting to changes in demand, and resourcing efficiently, particularly in terms of supplying the correct number of instructors (Harrington et al., 2017).

These results prompted the Air Force to explore potential initiatives that would help improve pipeline efficiency. The Air Force asked PAF to conduct additional research into several specific areas. In particular, AETC requested that we explore private-sector models that might offer insights and best practices upon which AETC could draw, in particular how colleges and universities right-size their instructor corps and best practices associated with supply chain management. In addition, AETC asked us to explore one potential solution—a flexible instructor pool—to address the challenging problem of rapidly sizing the instructor corps to respond to changing requirements that occur late in the planning process.[1] The intent of this research was to examine a series of approaches that AETC might consider as it reevaluates some of the foundations of its technical training processes.

More specifically, we evaluated the following topics. Each is addressed in a separate part of this report.

- From AETC's perspective, **colleges and universities** share many of the same attributes with Air Force technical training: They need resources (instructors, facilities, training devices, base operating support), and they must provide for an ever-changing number of students and types of courses required. Because of these similarities, AETC is looking for opportunities to improve their processes by learning from the experiences of colleges and universities in managing changing demands for education. Part II of this report discusses concepts employed by colleges and universities that may be applicable to AETC's training environment.
- RAND's work in FY 2016 suggested that technical training pipelines could benefit from analytical methods used in **supply chain management**. We suggested that decisionmaking might be better informed by a more-structured approach to planning and to providing resources for technical training pipelines—and that doing so would increase responsiveness to changes in training demands (such as adjusting for changes in the year of execution). AETC agreed that looking at technical training pipelines in a more systematic way (backed up by the extensive body of research available for supply chains) could make the pipelines more agile. Part III discusses supply chain management.
- AETC recognizes that resource limitations directly affect the agility of the training pipeline. Direction from Headquarters Air Force (the Air Force corporate level) does not allow AETC to carry excess capacity (especially extra instructors). Therefore, when demand for graduates increases over and above what was originally planned for, it takes a long time for resources to be allocated. AETC asked PAF to look at options to more rapidly access resources. An early consideration was a "working-capital-fund-like" approach, which did not end up achieving the desired aim. Instead, the team considered ways to create a **flexible pool of instructors** that could be accessed when required for unplanned increases in training requirements (either in the aggregate or for individual Air Force Specialty Codes [AFSCs], which is discussed in Part IV.

---

[1] The process to plan for the number of students who will enter and graduate technical training in any given year begins three years before the beginning of the FY (referred to as the "year of execution") and is supposed to be complete one year prior to the year of execution to allow time for schoolhouses to identify and bring on the requisite number of instructors based on student requirements. In recent years, changes have been made to this plan closer to and sometimes during the year of execution, which does not give schoolhouses the time required to bring on new instructors. Often, these changes involve small numbers of students to any given pipeline and do not require instructor changes. However, during FY 2016, an unexpected, significant, and last-minute congressionally mandated increase in Air Force end strength strained AETC's capacity beyond its limits and highlighted the lack of flexibility in the system to react on short notice. Although it is not clear whether large changes in training requirements will occur frequently in the future, interviewees expected last-minute changes to become the norm, prompting AETC to consider ways to improve flexibility rather than continue to operate with a system where flexibility is constrained by long planning lead times (Harrington et al., 2017).

Together, the topics in this report offer options and concepts that the Air Force could adopt to improve flexibility and responsiveness in the technical training pipeline. In the remainder of this report, we not only identify concepts of interest but also explain how the Air Force might integrate these concepts into its current approach to managing the nonrated technical training pipeline.

Part II. What Can the Air Force Learn from Higher Education Institutions About Meeting Challenges in Technical Training?

Matthew Walsh, Daniel Ibarra, Charles A. Goldman, Rita Karam, and Kathleen Reedy

2. Introduction and Overview of American Community Colleges

The Air Force manages a complex technical training enterprise that encompasses 379 Air Force Specialty Codes (AFSCs) and 887 individual courses. Air Education and Training Command (AETC), the parent organization for both enlisted and officer initial skills training, is charged with transforming civilians into airmen and delivering technical training to airmen in their respective specialties. One of the major challenges that AETC has faced in providing this initial training is rapid and sometimes sharp changes in demand for technical training graduates. The Air Force produces a Program Guidance Letter (PGL) that sets the number of graduates based on calculations of congressionally mandated end strength, retention rates, and specific career field needs. In theory, this letter is drafted three years before the beginning of the FY, when it is to be executed and finalized at least 12 months prior.

However, as the 2017 RAND study (Harrington et al.) highlighted, in recent years, this PGL has been finalized very late, there have been wide swings in end strength size, and changes are routinely being made to the PGL after the execution year actually starts. For some career fields, these changes in demand have been slight (only one or two students), but in other cases, the changes have involved hundreds of students. Harrington et al. (2017) showed that the AETC did not have the flexibility to respond rapidly to such changes in demand. This section attempts to explore how civilian institutions respond to similar fluctuations in demand and identify lessons that are applicable to AETC's processes.

In the civilian world, postsecondary education institutions in the United States have a similar mission: to educate learners and provide them with the skills needed to contribute to the workforce. The comparison with AETC is particularly clear in the case of community colleges, many of which grant certifications in areas of technical training.

Like AETC, postsecondary education institutions must meet dynamic and unforeseen student and employer demands using limited resources. AETC must provide quality training to a designated number of airmen in each AFSC despite diminishing budgets and resource constraints. Postsecondary institutions must also increasingly provide a high-quality education while controlling tuition costs.

Notably, both the education sector and AETC have been hit with funding challenges. Between 2008 and 2013, the Air Force budget declined 12 percent in real terms and is still recovering from the effects of sequestration (Williams, 2012). The Air Force has responded to tighter budgets and less resources in the training environment by taking on more risk in areas such as facilities and infrastructure. Similarly, during those same years, public universities and colleges saw an average of 28-percent reduction in state funding (Lapovsky, 2013), which is significant, given that state and federal funding largely subsidizes educational costs, representing about 44 percent of revenue in public nonprofit institutions (McFarland et al., 2017). Both educational institutions and AETC have had to find ways to improve efficiencies without the promise of additional resources.

Although the two sets of institutions are not identical, we examine practices from U.S. higher education to see what lessons can be applied to help the nonrated technical training enterprise become a more efficient organization. We describe four ways that postsecondary institutions attempt to be efficient and agile and consider potential benefits and barriers toward applying those practices to Air Force technical training:

1. how colleges outsource and automate services ancillary to their core mission
2. how colleges anticipate changes in workforce demand
3. how colleges adjust capacity in response to changes in demand
4. alternative methods for instructional delivery and when they are likely to be effective.

To address these topics, we used information from previous RAND higher education studies, additional reviews of published literature, and discussions with community college and Air Force officials working in these areas.

Overview of American Community Colleges

The U.S. community college system consists of about 1,108 schools located across the country, serving some 12 million students (American Association of Community Colleges, 2017). Community colleges generally offer lower-cost options for students to pursue higher education, although costs have been increasing markedly throughout higher education.

Community colleges offer two types of learning tracks: academic and vocational. The academic track allows students to take courses that fulfill general education and prerequisite requirements. Students in the academic track may choose to transfer to a four-year institution to pursue a bachelor's degree. The vocational track allows students to take courses focused on a specific trade or occupation and culminates in a certificate or an applied associate's degree.

Community colleges represent one of five major types of postsecondary institutions in the United States: the others are research universities, comprehensive universities, liberal arts colleges, and for-profit institutions. Collectively, these comprise more than 4,500 different schools serving more than 20 million students (National Center for Education Statistics, undated). Each type of institution has distinct aims and characteristics, yet all must deal, in some way, with the challenges of anticipating student and employer demands and providing affordable education.

In response to fiscal constraints and the changing needs of a growing population of nontraditional students, colleges have sought innovative solutions to effectively and efficiently accomplish their missions. Some of these solutions have arisen from the explosion of new educational technologies. Information technology (IT) systems have transformed how colleges deliver content, communicate with students, track their progress, and provide administrative and student support services. For example, a vast and growing number of students access courses online, allowing them to complete degrees despite geographical and time constraints that prevent them from attending classes in a traditional face-to-face setting (Allen and Seaman, 2011). Other solutions have arisen from findings in the learning sciences. For example, research supports the efficacy of a competency-based educational model in which credit is granted for demonstrating mastery of content rather than completing a prescribed number of credit hours (Means et al., 2010; see Part II, Section 6, on alternative delivery methods for more detail on this research). Colleges have increasingly adopted elements of competency-based education to reduce costs while improving learning outcomes.

There are a number of differences between postsecondary education and AETC technical training that limit the extent to which they can be compared. Colleges can impose their own limits on accessions and can choose to deny individuals admissions, whereas AETC must produce the number of graduates directed by Air Force headquarters. Relatedly, colleges are not required to produce a certain number of graduates, and while it reflects poorly on a school to have a low graduation rate, completing the curriculum is ultimately the responsibility of the students, not the schools. Schools also have the advantage of not needing to repurpose their technical workforce to provide instructors, whereas, for some of its workforce, AETC must remove airmen from operational units to assign them to schoolhouses as instructors.[2] Additionally, schools can more readily expand and reduce their workforce—for example, by requiring existing instructors to teach additional courses and, in some cases, with additional pay or by making use of adjunct instructors on an as-needed basis. Adjuncts provide flexibility in responding to changing demands because although the timelines of the hiring process can vary widely by program, once vetted, adjuncts have high retention rates and are only employed when there is demand (Magda, Poulin, and Clinefelter, 2015).

Uniformed instructors at AETC, by contrast, are on what are called "controlled tours," meaning that they are in place for three to four years regardless of changes in demand. As we learned from our interviews, government civilian instructors are often retained for even longer, as they provide consistency, and there is pressure to retain civilians within government service barring ethical violations or formal reductions in force. Course-wise, AETC courses are somewhat less flexible in their ordering (although, with a push to move to modularized, self-paced training in some career fields currently being piloted in AETC, this may change). They also do not have predictable, standardized semesters, although their schedules are usually set a year in advance. Notwithstanding these differences, there are still enough similarities between postsecondary education and AETC technical training to make it valuable to explore whether methods and processes that the civilian education system employs to meet the challenges of managing student flow and instructor resources are applicable to the technical training pipeline.

---

[2] AETC also has civilian and contract instructors that it uses to varying degrees across different schoolhouses.

3. Outsourcing and Automating Services Ancillary to the Core Mission

By the early 2000s, approximately 95 percent of educational institutions reported outsourcing at least one nonacademic or administrative function (Bushman and Dean, 2005). The percentage of institutions employing outsourcing and the range of outsourced services continues to increase. Privatization of non-mission-critical functions is seen as a viable—and necessary—step toward controlling higher education costs.

Institutions consider outsourcing for myriad reasons (Kerekes, 2010). Cost reduction is the single most important factor. Vendors can provide new technology and can capture economy of scale in a way that a single institution cannot, allowing them to deliver services for less. Additionally, outsourcing reallocates institutional resources by eliminating the need for large capital investments associated with the performance of certain functions. Outsourcing may also improve the quality of service, owing to the vendor's expertise in that area or contractual mandates regarding service standards. Finally, outsourcing may reduce staffing requirements, minimizing long-term employment-related costs.

With these benefits comes worry about the privatization of educational services (Bushman and Dean, 2005). Concerns include navigating existing labor agreements, maintaining quality, ensuring protection of institutional data and confidential student records, and developing and retaining internal expertise in new technology skills. These can be addressed, in part, by developing a process for evaluating and selecting vendors and ensuring appropriate oversight of vendors after awarding contracts.

Outsourcing typically involves services and functions that are not central to the core mission of education. Indeed, the most commonly outsourced functions in the early 2000s included food services (75 percent), vending (63 percent), student bookstores (46 percent), custodial work (26 percent), and laundry (21 percent) (Bushman and Dean, 2005). However, institutions are beginning to rely on private companies to perform central functions. For instance, online courses provide an important source of revenue to institutions and allow them to deliver content to more students (Karam et al., 2017). But many institutions lack the resources and expertise to develop and deliver online programs. To do so, they may form partnerships with private companies that have technology and funds to implement online programs and marketing expertise to advertise them (Karam et al., 2017). When outsourcing services, and particularly those central to the core mission of education, institutions must consider whether the values of the vendor are aligned with their own.

An alternate approach to reducing process costs is to form consortia (Gose, 2017; Smith et al., 1999). By working together, colleges can capture greater economies of scale, defray capital investments required of any single institution, improve services by exchanging best practices and processes, and share staff. An attractive feature of this approach, compared with outsourcing, is that colleges share the core mission of education.

A final way to reduce operating costs is to invest in automated solutions for administrative processes, student support services, and IT functions, sometimes called *student information systems*. Companies such as Pearson, Kaplan, and Blackboard provide technology approaches for higher education institutions to manage relationships with their students. Services provided by these firms include online program management, admissions, student recruitment/retention, marketing, online course delivery, IT help desk solutions, and registrar functions such as graduation checks and course enrollment. Students can access services using web-based interfaces. Faculty and staff can also use these services to view student data and performance metrics, identify students who are at risk academically, and communicate with students. The degree to which colleges use these services varies, with most using services such as learning management software and online course delivery systems, and some also electing to automate registrar and administrative functions.

Although administrative technology can reduce educational expenses, it may come with significant upfront costs associated with purchasing the software and computing infrastructure and training faculty, staff, and students to use the system (Manacapilli et al., 2011; Schmidtlein and Taylor, 2000). Technology may also come with sustained costs associated with paying annual licensing fees and providing ongoing technical support. In nonmonetary terms, the choice of a vendor and product may limit an institution's flexibility to modify or upgrade technology in the future. Finally, the introduction of educational technology raises concerns related to data ownership (i.e., students, institutions, or third-party vendors) along with data security (Fitzgerald, 2016).

Outsourcing or Contracting: Software Improvements and Automation in Air Force Technical Training

The two approaches just described—outsourcing (or, more correctly, contracting) and automating business processes—could enhance the efficiency and cost effectiveness of Air Force technical training. The Air Force, along with all other U.S. Department of Defense (DoD) agencies, outsources ancillary functions such as depot maintenance, military family housing, base operations support, ground maintenance, and transportation (Rendon, 2001). A 2015 survey of federal managers found that most rated their experience with outsourcing between satisfactory and good (Government Business Council, 2015). In the context of technical training, AETC could outsource additional functions related to instructional design and delivery.[3]

---

[3] We use the term *outsourcing* here because that is how it is referred to by educational institutes. Various policies and regulations are subject to change that outline the acceptable methods for outsourcing or contracting functions in DoD. Before making major changes, AETC should ensure that it complies with these policies and regulations (e.g., FederalSoup Staff, 2016).

New educational technologies have the potential to expand online and distributed learning (described in Section 6 of this part), specifically with respect to elements of technical training that can be delivered online. To realize this potential, staff must possess technical expertise in the software and computing systems used to deliver course content. In addition, they must possess pedagogical skills to develop course content that is consistent with the affordances and limitations of new and different formats. Given the limited training that Air Force instructors receive, it is unlikely that individual instructors will have the time to develop the requisite knowledge and skills to quickly and easily build and update courseware in such platforms. This problem is not unique to Air Force instructors—instructors in higher education also require additional training and ongoing support to effectively leverage new educational technologies (Devaux et al., 2017; Manacapilli et al., 2011). Contracting elements of course development and upgrades would eliminate the need to provide extensive additional training to Air Force instructors, while still allowing AETC to reap the benefits of new educational technologies. To ensure that materials provide comprehensive coverage of topic areas and meet Air Force training standards, contractors can develop products in collaboration with Air Force instructors.

Educational institutions face the daunting task of accepting large numbers of students, registering and scheduling them in courses, and auditing their progress toward degree requirements, often seeking outside help or software to do so. Similarly, managing the flow of students through Air Force technical training pipelines is complicated also by a range of factors (Harrington et al., 2017). Students arrive at schoolhouses from multiple sources including Basic Military Training (BMT), colleges and universities, the Air Force Academy, and home units (in the case of reservists). Despite heavy planning, schoolhouses report that arrival dates at technical training remain inexact, and student start dates may vary (particularly for guard and reserve students). Some training squadrons have reported that students will arrive at schoolhouses without the necessary security clearances or waivers to begin training (because of broader DoD clearance processing slowdowns). Finally, during training, students may fail sections, forcing them to repeat earlier sections of the course. All of these make real-time managing and scheduling a complex process in the technical training pipeline.

AETC could, then, similarly explore outsourcing administrative functions (such as scheduling and tracking student progress) or contracting with organizations to develop and update Air Force–specific student information systems and software for tracking and automating functions (e.g., scheduling) to leverage their expertise to improve the efficiency of the current process.

4. Anticipating Changes in Workforce Demand

To adequately prepare graduates to join regional and national workforces, colleges must track labor demand. Additionally, they must anticipate changes as historically stable industries fade and newer industries emerge. In response to—or in advance of—these changes, colleges must evaluate the set of degree programs offered and the curricula that make up each program to gauge both the long-term trends and the likely short-term changes in requirements for faculty numbers.

Educational institutions use a range of techniques to track and anticipate changes in workforce demand, such as extrapolating from historical data. Sources such as the Bureau of Labor and Statistics and the American Community Survey provide census data on national workforce trends, while sources such as the Texas Workforce Commission provide data on regional trends. Institutions use these databases to gauge career field demand from previous years and to predict changes in demand (Karam et al., 2017). Relatedly, institutions can use more-sophisticated modeling methodologies, such as stock or flow models, to account for the composition of the workforce in addition to labor market demand when predicting future needs (Goldman et al., 2015).

Educational institutions also talk with employers and industry experts. In a recent survey of postsecondary Texas institutions, interviewees typically reported placing greater weight on employers' opinions about future needs than on census data (Karam et al., 2017). In addition to communicating informally with employers, many institutions and departments form advisory boards made up of industry experts. Advisory boards continually assess market needs and future trends and provide institutions with actionable information. Lastly, state and local governments sometimes initiate partnerships between employers, schools, and colleges. For example, the California Career Pathways Trust invests in such partnerships to better prepare California students for the 21st-century workplace. Internationally, European countries have enacted legislation to ensure that public and private sectors work collaboratively with schools to improve technical and vocational education and training (Constant et al., 2014).

Educational institutions use a range of other techniques to gauge workforce demand. They may interview program alumni and analyze graduate placement rates (Constant et al., 2014; Karam et al., 2017). They may also track economic indicators such as "hot jobs," rapidly growing occupations, substantial wage changes, and vacancy rates (Goldman et al., 2015). Moreover, institutions may analyze job advertisements (Constant et al., 2014). Finally, they may hire consulting companies to estimate workforce needs (Karam et al., 2017).

Two methods for assessing workforce demand—talking with employers and alumni—may provide additional information about the competencies graduates will need in a career field

(Constant et al., 2014). In addition to helping institutions decide which programs to offer, these methods can help institutions decide what to cover within a program.

Estimating Workforce Demand in Air Force Technical Training

Estimating workforce demand for Air Force technical training should, *in principle*, be easier than estimating demand in postsecondary education for three reasons. First, Air Force technical training programs map directly to Air Force career fields, whereas college degrees may lead to work in one or more areas. Consequently, once the required number of new technicians in an Air Force career field is specified, Air Education and Training Command (AETC) knows the required number of accessions that will be incoming to meet that demand. Second, educational and operational managers in the Air Force belong to the same overarching organization, whereas educational institutions and public and private companies are separate entities with different values and motives. Third, the Air Force can capture and record detailed information about its workforce. This information can enable formal approaches to workforce modeling (Goldman et al., 2015). In practice, communication between various echelons within the Air Force is not always timely, and methods of communicating demand signals between different Air Force Specialty Codes (AFSCs) are not always consistent (Harrington et al., 2017).

To achieve more efficient and accurate planning, AETC can leverage two approaches from higher education: formal modeling and collaboration with operational managers. Stock modeling is an example of a formal approach to degree program planning in higher education (Goldman et al., 2015). This methodology involves four steps: (1) project total demand for workers by occupation; (2) compare projected demand with projected supply; (3) link occupations to postsecondary degree programs; and (4) compare current and projected capacity of postsecondary programs to meet projected demand. Congress sets the aggregate end strength; then, using stated manpower requirements and retention rates, the demand for each Air Force occupation (i.e., AFSC) is determined. By design, technical training pipelines are aligned with one AFSC. As such, inputs to Steps 1 and 3 of the stock model are given. Using detailed information about the current composition of the Air Force workforce (e.g., age, skill-level, time-in-service), AETC could collaborate with Air Force, Manpower, Personnel and Services (AF/A1) to pursue new approaches to better predict projected supply (Step 2). Additionally, by recording training pipeline performance data, AETC can more accurately predict projected capacity by technical training program (Step 4). Because the types of data needed for workforce modeling approaches are available, the Air Force should consider whether these approaches can add value beyond the Air Force skills-projection model and career field sustainment analysis currently used.

In addition to formal modeling, the Air Force should work to ensure greater communication between echelons of Air Force technical training, as colleges and universities have attempted to do with employers. Harrington et al. (2017) noted a lack of readily available information to support execution planning. For example, changes to the PGL, the document specifying annual requirements for the number of trained personnel in each AFSC, may be communicated to schoolhouses within only months of the start of the fiscal year. This limits their ability to respond to changes in demand. Conversely, information about schoolhouse capacity and throughput is not visible to headquarters AETC. A centralized database and a formalized process for communicating across echelons of Air Force technical training would improve estimation and responsiveness to changes in workforce demand both in the short-term for flexibility but also for long-term planning about requirements for numbers of teaching staff.

5. Adjusting Capacity in Response to Changes in Demand

Academic departments must anticipate and adapt to changes in registration, while colleges must anticipate and adapt to the number of freshmen accepting admission offers. Too large of an incoming class may exceed the institution's educational and physical capacity and force them to limit enrollment (Hussain, 2017). How, then, do colleges and departments adjust to unpredictable and dynamic student demand?

Colleges largely accommodate variable demand by expanding and contracting the teaching force. This is often particularly necessary for lower-level courses, which often have the highest demand, as enrollment is generally higher among freshmen and sophomores and general education courses are required for many degrees (Guidry, 2015). Hiring tenure-track faculty is time consuming, expensive, and often impractical. Many colleges rely instead on contingent faculty to teach a significant portion of their courses. Contingent faculty includes both full- and part-time nontenure-track faculty. These positions are referred to as *lecturers*, *staff*, and, in the case of part-time faculty, *adjuncts*. The common characteristic is that colleges make little or no long-term commitment to them. Adjuncts now make up a majority of university faculty: By 2013, adjuncts made up 50 percent of public research university faculty, 57 percent of private research university faculty, and 83 percent of public community college faculty (Hurlburt and McGarrah, 2016).

Educational institutions primarily hire adjuncts because they are readily available and inexpensive. Additionally, adjuncts may bring unique and valuable technical experience from concurrent private-sector employment to the classroom (Bettinger and Long, 2010), however, they may have little experience in education. Additionally, because adjuncts receive such low pay, high-quality instructors may not apply for adjunct positions. More generally, the practice of hiring a largely transient workforce may undermine the cohesion of an academic department. Adjuncts have limited commitment to each campus, so they cannot form the core of a program or department; those responsibilities must be shouldered instead by the (perhaps few) full-time faculty members.

Expanding the scope beyond community colleges, colleges and universities use teaching assistants (TAs) to expand their workforce. TAs are typically graduate students or advanced undergraduates who have previously completed the course. They perform a range of functions, such as leading small laboratory or recitation sections that meet separately from full class lectures, grading homework and exams, preparing assignments, and holding office hours. TAs are not intended to supplant instructors. Rather, by performing support functions, they allow instructors to take on more or larger classes.

A final response to increased student demand is to become more selective in enrollment. For example, the growing demand for construction management graduates during the mid-2000s caused a surge in applicants, which exceeded the capacity of accredited construction management baccalaureate programs in the United States. When asked how they managed the surge, colleges most frequently cited limiting enrollment (24 percent of institutions surveyed), followed by hiring adjuncts to increase the number of courses offered and the program's capacity (20 percent) (Christofferson, Wynn, and Newitt, 2006). Although limiting enrollment does not address student demand, admitting more qualified individuals may allow a program to graduate a larger percentage of students, increasing the quality of the institution.

Using Adjunct Faculty to Adjust Air Force Technical Training Capacity

We held numerous discussions with Air Force stakeholders and subject-matter experts within the technical training pipeline as part of its FY 2016 research on the technical training pipeline. A common theme from these discussions was the need for more instructors at schoolhouses that were operating at or close to capacity. In addition, instructors had difficulty balancing administrative burdens with instruction time and curriculum development. Moreover, the ability to surge to meet increases in throughput and changes in end strength was frequently accomplished by piling more demands on the existing instructor corps.

Air Education and Training Command (AETC) currently requires that instructors be nominated through the Developmental Special Duty (DSD) process, which has fairly stringent instructor-selection requirements. In meetings with staff and civilian instructors at schoolhouses, we heard that DSD has increased the quality of the instructor corps within the technical and basic training environments, but it has decreased the timeliness of assigning instructors to vacant positions. The process of identifying an instructor through DSD, reassigning them to a schoolhouse, and qualifying them to teach can take a year or more. Part of the delay relates to the fact that selection boards only meet biannually to nominate new instructors. If Air Force demand for airmen in certain Air Force Specialty Codes (AFSCs) surges, the lengthy DSD process—and the time to reassign and train instructors—prevents schoolhouses from quickly responding in kind.

How can AETC create a more-adaptive instructor workforce, as colleges and universities have done, to better respond to changes in demand? This could be done in two ways. First, qualify more instructors through the DSD process than are immediately needed. Creating an instructor reserve would allow AETC to respond to surges without needing to wait for the next DSD board to meet, an option that, as we learned from discussions with stakeholders in these organizations, the 2AF and AETC are already considering. This approach would reduce, but not eliminate, the time needed to hire a full-time instructor (see Part III for more on the made to order, made to stock, and assemble to order options).

Second, AETC could create a flexible "adjunct" instructor corps to respond to surges, as colleges have done with adjunct faculty. The adjunct instructor corps could be made up of former instructors and guard and reserve members. Former instructors provide a qualified and experienced pool of adjuncts. AETC could implement processes to ensure that former instructors receive periodic refresher training to remain ready to teach and undergo rapid requalification prior to returning to the classroom (see Part IV for further analysis of these options and their relative advantages and disadvantages).

An adjunct instructor corps could be employed in two ways, depending on the course and AFSC: through temporary duty (TDY) assists, where instructors return to the schoolhouse and teach in-person for periods of time (either one course's worth or up to full-time, see Part IV for various other options), and remotely, where instructors remain at their home stations and teach using synchronous or asynchronous online learning platforms (described later in Section 6 of this part). Training squadrons could decide which path to use based on considerations, such as the length of the course, the burden on operational units of losing personnel to instructor TDY assignments, whether surge capacity would need to be for the full length of the course or just for a portion of it, and the need for some courses to be taught in person because they require hands-on instruction or must be conducted in classified facilities. Perhaps the biggest obstacle to leveraging former instructors still on active duty is managing conflict with their current assignments. Operational units may resist giving up personnel to TDY assists and doing so may increase the operational load at those units. Allowing instructors to teach during personal time from their home stations and for supplemental pay may be a more acceptable option. Operational units would retain operational capacity, and instructors would voluntarily participate for supplemental pay. Of course, this would only work for classes that could be taught online (e.g., courses that do not include classified systems or material).

Guard and reserve members who possess technical knowledge and instructor training could also serve as adjunct faculty. Recently, Air Force Reserve Command (AFRC) announced a new capability within an IT system called ARCNet, which affords Guard and reserve members the ability to upload both civilian and military resumes. Through ARCNet, National Guard and reserve members can track their readiness, training requirements, and view personnel and career data. The new Volunteer Reserve System (VRS) will allow requisitions to search reserve members' skills across their military and civilian careers, which will aid in identifying individuals who may be a good fit for critical jobs. AETC could take advantage of this system to identify potential instructors for technical training who have relevant experience from their Air Force or civilian jobs. In essence, using the VRS gives AETC another mechanism to expand the pool of potential instructors, rather than relying purely on the military's direct capabilities.

The biggest obstacle to leveraging guard and reserve members in this way is securing the proper funding to allow schoolhouses to use them as instructors without losing active-duty billets. Using guard and reserve members would also require balancing military commitments with civilian employers. Lastly, and as with an active-duty adjunct corps, guard and reserve members would need to obtain and maintain instructor qualifications.

Historically, using reservists as teaching staff was not cost effective because they were intended as replacements for active-duty instructors and came with a one-year commitment. We stress, therefore, that reservists should be used to supplement full-time staff during periods of unexpected surges, either through TDY assists or teaching remotely with online platforms. These options come with some of the same costs as retaining former active-duty instructors, but they create new opportunities to flexibly expand the instructor workforce (discussed further in Part IV).

6. Alternative Methods for Instructional Delivery

Colleges have explored alternative instructional methods to enhance or replace a traditional pedagogical model in which instructors deliver lectures to students in a face-to-face setting. The potential benefits of alternative instructional methods include improved student outcomes, reduced educational costs, and greater accessibility of education to nontraditional learners. Some institutions have implemented these approaches in one or a small number of courses, and some have built full degree programs around them.

In this section, we describe three alternative instructional methods: online courses, blended courses, and competency-based education. We review studies that examine their effectiveness, and we discuss factors that AETC must consider before adopting them.

Online and Blended Learning

Colleges have increasingly created online programs because of their potential to allow students to flexibly access instruction and content at any time and from any place. Reasons for implementing and expanding online programs include (1) ease of providing education to students who cannot or choose not to attend classes in a traditional face-to-face setting, (2) reuse of educational material leading to reduced cost per student, and (3) scalability, enabling instructors to teach larger classes without negatively affecting educational quality (Means et al., 2010).

Online courses include but go beyond the computer-based training commonly used to deliver required training en masse across the military, in which the learner does not interact with an instructor or peers. Information technologies allow students to communicate with one another and with an instructor to create a much more engaging and stimulating environment. Different information technologies support different models of online learning. One model uses asynchronous communication tools (e.g., email, discussion boards) to allow students to share ideas and ask questions at their convenience. Another model uses synchronous communication tools (e.g., telephone, videoconferencing, chat) to allow students to interact in real time. Courses often combine multiple forms of asynchronous and synchronous communication to support different learning experiences.

In addition to using online learning as an *alternative* to face-to-face instruction, colleges use online learning to *enhance* face-to-face instruction. This is called *blended learning*. Blended courses combine classroom and online learning and activities. Courses vary in terms of the allocation of time between classroom and online learning, and the types of activities delivered in each. These design choices are based on characteristics of the learners, of the instructor, and of the desired learning outcomes (University of Washington, 2013).

Potential benefits of blended learning overlap with those of online learning: increased ease of access, reduced educational costs, and increased student capacity. These benefits arise from the fact that students may come to campus less often and remain for less time. In addition to these benefits, the blended format allows instructors to combine strengths of face-to-face learning with those of online learning. Online and face-to-face learning can be combined in novel ways. For example:

- Flipped classrooms: Students watch video lectures on their own time and at their own pace prior to class meetings. They then practice methods covered in lectures alone or in groups during class meetings and with instructor support (Steele et al., 2014).
- Simulation: Students attend class meetings. They then practice applying knowledge and skills in computer-based simulations that emulate authentic problems (McGrath et al., 2016).
- Computer-based tutors: Students attend class meetings. They then review and practice content using computer-based, "intelligent" tutors. Tutors advance when the student has demonstrated mastery, tutors provide feedback, and tutors tailor content to the needs of the student (Koedinger et al., 1997).

Online and blended learning have tremendous potential to positively affect education, however, they also pose several challenges. First, the upfront costs of purchasing software and computing infrastructures, along with annual licensing fees, may be prohibitive (Steele et al., 2014). Even if students are not provided with personal computers, institutions may lack the computer hardware on-site to allow students to access online instructional tools. Second, building new course content for online delivery is costly and time consuming (Manacapilli et al., 2011). Third, instructors need support to use new technology and to effectively redesign and deliver courses (Devaux et al., 2017). Students may also need support to use new technology. Fourth, online learning places greater responsibility on students. Success in online courses, especially asynchronous ones, depends largely on students' self-motivation, since there are no face-to-face interactions with instructors. The absence of an instructor can make it easier for students to participate minimally or to completely withdraw from the course (Cull, Reed, and Karin, 2010). This concern is somewhat mitigated in blended courses, which include face-to-face interactions.

Competency-Based Education

Online and blended learning emerged from the introduction of new educational technologies. Competency-based education, in contrast, reflects long-held ideas from the progressive education movement; namely, that students should develop broadly applicable "competencies" rather than learning narrow, content-specific standards (Steele et al., 2014). Competency-based education is defined in various ways, yet most sources agree on one central feature: Competency-based education specifies what a student should know and be able to do upon completing a course (i.e., "competencies"), and it awards course credit when a student demonstrates mastery of those competencies (Daugherty, Davis, and Miller, 2015). Other important features of competency-based learning are measurable learning objectives, constructive assessments, and timely and personalized support based on student needs (Sturgis, 2015).

Competency-based courses do not have a required set of materials that students must complete (Daugherty, Davis, and Miller, 2015). Instead, they provide a range of materials to students. Students may be allowed to skip material based on previous work experience or the results of pre-course assessments. Students may also be allowed to choose from two or more activities to satisfy a learning objective. Thus, competency-based learning allows for variation in *what* is covered and in *how* it is covered. Courses rely exclusively on a set of assessments to determine mastery. Students can take the assessments as soon as they have mastered a competency, allowing for individual variation in pace.

Some programs even award credit for previous work experience and allow students to test out of content based on prior learning assessments (Brigham and Klein-Collins, 2010; Freed and Mollick, 2010). This flexible pacing and reduced duplication of effort allows students to complete competency-based programs more efficiently and at a lower cost (Daugherty, Davis, and Miller, 2015; Miller and Davis, 2015). In addition to reducing costs, competency-based education can potentially align educational programs with employers' needs. Most programs focus on a set of competencies that are identified in conjunction with employers.

To successfully implement competency-based programs, institutions must overcome several challenges (Daugherty, Davis, and Miller, 2015). First, the movement away from credit hours, the potential for variable course lengths by individual, and the introduction of new grading mechanisms may be incompatible with existing administrative software and processes. Second, competency-based programs may be incompatible with regional accreditors. Third, competency-based education transforms the instructor's role. Less emphasis is placed on lecturing, and more emphasis is placed on compiling and creating course materials, guiding students, providing instructional support, and administering and interpreting assessments. Instructors may require support and training to adjust to this new role. Fourth, creating new content for competency-based programs is costly and time consuming. Finally, as with online and blended learning, competency-based learning places greater responsibility on students. Instructors must monitor students for early signs of problems. This can be facilitated by using data systems that pinpoint problems and issues before they arise and that provide supplemental information to students and instructors (Miller and Davis, 2015).

Effectiveness of Alternative Instruction Methods

Online learning, blended learning, and competency-based education have been widely adopted throughout higher education, as well as in K–12 education and cooperate training. For example:

- In a recent survey of more than 2,500 colleges and universities, 65 percent of reporting institutions considered *online learning* to be a critical part of their long-term strategy. The study also found that 31 percent of all students were currently taking at least one online class (Allen and Seaman, 2011).
- In another survey of higher education technology professionals, 90 percent of respondents said their institution encourages instructors to use creative teaching practices through *blended learning* (Center for Digital Education, 2012).
- The Carnegie Foundation for the Advancement of Teaching (2014) found that 29 states allowed K–12 districts to choose *competency-based* crediting if they wished, while another ten states had policies that granted seat-time waivers under certain circumstances. In terms of postsecondary education, the majority of the general public surveyed in a 2012 Gallup poll indicated that students should receive college credit for skills learned outside the classroom (87 percent agreed), and that students should be able to earn course credit in less time than the standard academic semester if they can demonstrate mastery of all the content sooner (70 percent agreed) (Lumina Foundation, 2013). Finally, competency-based approaches have a long history in higher education and are currently implemented in many postsecondary institutions (Klein-Collins, 2012).

These alternative approaches have gained increasing attention. But are they effective? With respect to online and blended learning, there is no consensus regarding their effectiveness compared with face-to-face instruction. The results of field studies and meta-analyses are inconsistent for two main reasons. First, few field studies involve rigorous experimental designs with random assignment or strong quasiexperimental designs (Means et al., 2010). Second, mode of delivery is only one of several variables that impacts learning outcomes. Other factors that must be considered include student motivation, course structure, instructor quality, course content, and quality of interaction between faculty and students.

In 2009, the U.S. Department of Education commissioned a meta-analysis to examine learning outcomes in online and blended courses compared with traditional face-to-face courses. Of the roughly 100 studies comparing online and face-to-face conditions between 1996 to 2008, only 45 met the definition of rigorous designs. Of those studies, five involved K–12 learners, and the remaining 40 involved learners in community colleges and four-year colleges. Findings from the meta-analysis indicated that online learning was about as effective as face-to-face learning, whereas blended learning was superior to face-to-face delivery. The meta-analysis further found that online courses that were collaborative or instructor-directed led to more positive learning outcomes than those that students completed independently. Finally, the meta-analysis highlighted effective features of online courses such as giving learners more control of their interactions with media, prompting them on their work, and requiring them to self-monitor their performance (Means et al., 2010; see also Sitzmann et al., 2006).

A more recent study compared online and face-to-face courses within the Washington State Community College System (Xu and Jaggars, 2013). The study used statistical methods to control for student characteristics and compared learning outcomes for students who took the class in the online format versus the face-to-face format. In contrast to the previous meta-analysis, student retention and end-of-semester grades were somewhat lower in the online condition. The study suggested that the negative results may have related to the academic unpreparedness of community college students, lack of extrinsic motivation, and weak instructional pedagogy, factors also shown to impact the success of online learning in the earlier meta-analysis.

As with studies of online and blended learning, empirical investigations of competency-based education have yielded mixed results. Anecdotally, efforts to undertake competency-based reforms in primary and secondary schools have produced promising results, including improved reading and math scores and increased graduation rates (for reviews, see Lewis et al., 2013; Steele et al., 2014). These results are merely suggestive, however, as the case studies lack suitable controls for comparison, and the ways in which schools implemented competency-based education were not operationalized.

Studies that have systematically investigated the effectiveness of competency-based education compared with other online and face-to-face formats in community colleges have produced mixed results. Yarnall, Means, and Wetzel (2016) found that adaptive (i.e., competency-based) courseware resulted in slightly higher grades and more positive student learning in select cases. Moving from a lecture-based model to one that incorporated adaptive technology and moving from nonadaptive to adaptive learning systems were associated with the most positive outcomes. Another recent study compared learning outcomes in three community colleges that adopted competency-based programs with a set of peer institutions. The effects of competency-based education varied across site (Person et al., 2016); programs that incorporated academic coaching produced the highest rates of student credentialing (Person et al., 2016). There are also numerous challenges with competency-based education, primarily in ensuring that assessments are reliable and targeting the appropriate knowledge and skills, that instructors are prepared to teach to this sort of rubric, and that such education is systemically integrated into and recognized by the rest of the institution. Such challenges need to be addressed prior to and throughout implementation of this type of education program (Klein-Collins, 2012, pp. 29–31).

A final consideration is whether competency-based programs reduce educational costs. Surveys and analysis of competency-based programs across institutions consistently show that these programs recoup upfront investments and reduce educational costs within their first few years (Desrochers and Staisloff, 2016; Yarnall, Means, and Wetzel, 2016).

Application of Alternative Instruction Methods to Air Force Technical Training

AETC must take several considerations into account before deciding to implement online learning, blended learning, and competency-based education. These considerations fall into four categories: student access, student motivation, instructor quality, and resources.

Student access. The Air Force is geographically dispersed. In addition, any education completed outside the period of formal technical training must be balanced with duty station responsibilities. As such, the Air Force requires "diverse training delivery methods that are available to Airmen anywhere anytime through a robust integration of technology" (AETC, 2008). The concept of distributed learning is not new to the Air Force. Many AFSCs allow airmen to complete Career Development Courses and on-the-job training remotely. The idea of technology-enhanced distributed learning is also not new to DoD. Since the DoD launched the Advanced Distributed Learning Initiative in 1997, use of distributed learning has increased dramatically, with service members and DoD civilians completing more than 10 million online courses annually (Wisher, 2011).

AETC may be able to integrate online and blended learning into initial skills training to deliver learning anywhere anytime. For example, some portion of technical training could be offloaded to the airman's initial duty station to increase schoolhouse capacity where they would be considered on TDY-in-place.[4] For training wings that share common resources at a base (e.g., classroom space, dormitories), making some training available in this format would additionally reduce pressure on other resources for AFSCs that would still require in-person training (see Part III's discussion on binding resources). However, doing so would first require demonstrating that the reduced cost of initial skills training offsets the greater cost of on-the-job training (Manacapilli et al., 2007), as well as ensuring that such a system is employed with due consideration as to how to handle any students who have to be released from the Air Force or reclassified for failing to complete training. Alternatively, blended learning could reshape the eight-hour instruction day by having students view lectures in multimedia formats before scheduled classroom sessions (i.e., flipped classroom). Instructors could then focus on active learning activities during class time. Additionally, recording lectures in advance would reduce lecture time, allowing instructors to teach more classes or larger classes.

Career Development Course training may be even more suited for online and blended learning. For instance, students could complete all or some portion of a course during a portion of their duty time at their home station. The distributed portion of the course could be followed by an abbreviated residential period to conduct capstone group exercises and related assessments.

Student motivation. Characteristics of the student, and of the relationship between the instructor and the student, are critical to the success of online and blended learning. The absence of face-to-face interactions in online courses may reduce student engagement and motivation. Further, the absence of face-to-face interactions may limit the instructor's ability to anticipate and address problems. For these reasons, online courses work best with students who are self-

---

[4] In such an event, AETC would have to develop policies about accountability and whether the gaining command or AETC will be responsible for these students.

motivated, have self-discipline, and know how to manage their time. Blended learning minimizes these concerns somewhat, as students spend time in face-to-face meetings as well.

Student motivation may be less of an issue in Air Force technical training for various reasons. College students are free to drop courses, whereas airmen in technical training are not. Additionally, owing to its alignment with Air Force career fields, students may be more likely to view technical training as valuable and authentic. Finally, we envision online learning as part of a blended curriculum, which includes face-to-face meetings.

Incorporating elements of competency-based education into online and blended learning can further enhance student motivation. Competency-based programs include pre-course assessments and ongoing assessments that allow students to advance immediately upon demonstrating mastery of a competency. By reducing time needlessly spent reviewing mastered material, competency-based programs can increase student motivation. Defining defensible metrics of prior mastery in Air Force technical training would be challenging. However, delivering assessments in an online format could allow students to complete aspects of the training curriculum in an accelerated manner.

Instructor quality. The mode of instruction does not determine educational effectiveness. Distributed learning and in-residence learning, when delivered properly, are both effective (Clark and Mayer, 2011). As such, it is critical that instructors teaching online courses for the first time receive pedagogical training and support to ensure that they are equipped with effective strategies. They should not simply convert face-to-face materials to an online format and use technology to deliver lectures in a traditional manner. Instead, instructors should use techniques that create an online environment that encourages collaboration, discussions, inquiry, and reflection (Straus et al., 2013).

Instructors must also assume new roles to deliver competency-based education (Daugherty, Davis, and Miller, 2015). In particular, they must learn to compile materials and guide students to meaningful exercises, provide students with instructional support, and administer assessments and certify results. The most time-consuming and challenging aspect of competency-based education is course design, however. Given some training and support, along with an existing, well-designed course, Air Force instructors could effectively deliver competency-based education. Again, this would require an upfront investment of time and expertise in course design.

Resources. Online programs have lower operating costs than traditional programs, and they minimize or eliminate many student expenses (e.g., commuting costs, student housing, textbooks and materials). As such, online programs may be more profitable for colleges (Desrochers and Staisloff, 2016; Yarnall, Means, and Wetzel, 2016), and more affordable for students (Karam et al., 2017; Straus et al., 2013). Yet these programs come with the significant upfront costs of purchasing software and equipment, training instructors to teach courses in the new format, and creating new content. The costs of hybrid programs may be even greater because of the dual-use of online and face-to-face instruction. However, given the large number of students that complete Air Force technical training, upfront investments in online and blended learning will likely be returned. More importantly, these investments may provide schoolhouses with greater capacity to respond to surges in demand.

Competency-based education may also be more economical. Individuals cross-training into a new AFSC may already possess requisite knowledge and skills. These individuals could potentially test out of early sections of courses and advance more quickly through later sections. As we learned in discussions with training managers at schoolhouses, there currently are options available to students to "proficiency advance" (i.e., test out of courses if they already know the material), but there is a great deal of pressure on students to not take this option. First, the risk is often considered too high, because if students fail, they acquire a failing mark on their records, making it safer to sit through the entire course. Second, the current institutional approach across AETC places a high degree of value on the importance of indoctrination into Air Force culture and values obtained by being the member of a cohort and progressing through the entire class with peers. Finally, training managers cautioned that, in the current system of planning and scheduling students in physical classes, having students that might join a class at any time would present challenges for schedulers and possibly increase the numbers of students waiting for training seats to open. In its current form, the constraint of moving through Air Force technical training as part of a class may undercut the benefits of competency-based learning. However, with policy changes to mitigate the risk of attempting proficiency advancement or when combined with a new system of online or hybrid learning, competency-based education has the potential to reduce training time and ultimately costs.

Final Considerations for Adopting Alternative Instruction Methods

Given the economical evidence along with suggestive findings regarding student outcomes, online and hybrid learning seem like valuable educational technologies worth adopting. The question is how AETC can best incorporate these technologies into Air Force technical training. Although they may have an obvious role in career development courses, which are often completed remotely, it is less clear how to leverage them in initial skills training. A blended approach might reduce instructor demand while allowing students to engage in authentic problems, simulation, and computer-based tutoring.

Competency-based education is largely compatible with Air Force technical training—perhaps more so than with K–12 or postsecondary education. This is because technical training is structured around a set of well-defined, occupation-relevant competencies learned in the classroom and demonstrated in exercises. Additionally, technical training pipelines are broken down into modules. Students only advance after demonstrating mastery of a module, typically by performing satisfactorily on a test or exercise. Although there are opportunities for "proficiency advancement" by testing out of a module, failing this test results in students receiving a failing grade on their record rather than just requiring them to take the module—thus creating a disincentive for students to attempt to test out. Removing the punitive measures and ensuring that assessments are thorough and validated to meet the course requirements—and doing so to the same or greater extent than current courses do—would create opportunities to move students through training at a much faster rate, saving time and money. Finally, technical training is not subject to educational accreditation across districts and states. AETC should continue to consider which practices from competency-based education can be used to enhance technical training.

Regardless of the alternative instruction methods selected, AETC will need to invest in new technology, instructor training, and curriculum development. In addition, any training change would require a concomitant assessment plan to ensure that learning outcomes are not diminished and the change is economically sensible and sustainable. This will require a detailed, qualitative assessment of current graduates in face-to-face courses to establish a baseline—and longitudinal tracking of students during the courses—immediately after graduation and within one to two years of arriving at their initial duty station. At the time of writing, AETC's cyber technical training courses have begun implementing pilot programs to test the appropriateness and efficacy of competency-based education and advancement in technical training. No feedback is available yet on their assessed effectiveness, but the results from these pilots will have implications for expanding such a model to other courses.

7. Conclusion

Colleges and universities face many of the same challenges as Air Force technical training and, therefore, some of the practices colleges have adopted to meet these challenges might also hold promise for the Air Force. The two main advantages that colleges have over AETC in terms of planning for and providing education is their ability to rapidly flex when demands change and to more readily adopt new methods of planning and instruction. The use of adjunct instructors combined with online and blended learning, in particular, allows these institutions to quickly adjust the size of their work force to respond to unexpected surges of students without having to make long-term commitments to additional staff. In addition, community colleges focused on technical degrees make a strong effort to work with the local community to identify educational needs and expected throughput in certain career fields. Finally, colleges and universities can also take advantage of outside expertise by outsourcing parts of their registrar functions and quickly adopting the latest technological advances to enhance the training and planning processes. Although there are differences between colleges and universities and the Air Force's training mission, AETC may benefit from adopting or adapting some of the practices currently being employed in post-secondary education, including:

- Planning and resourcing approaches
  - **Outsource or contract for administrative functions and tools**. Some colleges and universities turn to companies with expertise to either fully run or to help develop internal software to improve or automate functions such as scheduling and tracking student progress. The Air Force could consider leveraging similar expertise to either employ or adapt existing tools, processes, and software to the Air Force environment to help reduce the complexities of scheduling and tracking airmen's progress.
  - **Improve planning with modeling and improved communication**. Requirements for student numbers are driven by complex calculations at AF/A1 that are delivered as products to AETC. AETC could get ahead of some of the uncertainty in this planning system by coordinating with AF/A1, the 2AF, functional managers, and the training squadrons to develop models that may help assess levels of future need and therefore give AETC more lead time to plan in the long-term—not unlike how colleges and universities conduct environmental scans to determine future demand. Such efforts are currently hindered by inefficient communications in the technical training pipeline and perhaps could be alleviated by more consistently involving functional managers at the annual technical training planning conferences.
  - **Develop an adjunct-type model for the instructor corps.** This type of model will help provide additional capacity and support during periods of student surges without requiring AETC to bring on additional full-time instructors. Potential options to increase the pool would be TDY assists from former instructors, distance teaching provided by former instructors, and local, instructor-trained guard and reserve members. (Part IV contains a detailed assessment of options for developing a flexible

instructor pool.) Alternatives should be evaluated at the schoolhouse or by functional communities, as specific details such as course length and content affect which options are appropriate.

- Instructional approaches

    - **Explore the use of online or blended learning for initial skills training.** Incorporating online or blended learning into the curriculum offers two advantages. First, it enables adjunct instructors who are currently assigned elsewhere to instruct courses remotely outside of normal duty hours, increasing the pool of potential instructors. Second, it may allow for the students to proceed directly to their first duty station and continue to be in a student status for some or part of their initial skills training, reducing student capacity pressure at the schoolhouses and increasing shared resource capacity at training bases (see Part V, Section 16 for assessments of capacity flexibility).[5]

    - **Evaluate the use of competency-based learning.** Competency-based learning can be an effective educational approach that, when combined with online or hybrid learning, may decrease the instruction time students require before graduation. Currently, there are institutional barriers that limit the utility of such an approach. By making changes to some of those factors, AETC could effectively incorporate competency-based learning to reduce costs and time spent in training.

---

[5] Some interviewees emphasized the importance of technical training schools for continuing to "Blue" airmen (i.e. to continue to teach them the basics of Air Force culture after BMT). However, other interviewees noted that the length of time spent in technical training can vary widely between schoolhouses and that this cultural indoctrination continues into the first duty assignment anyhow, suggesting that such a point is less relevant.

Part III. A Supply Chain and Production Systems Approach to Programming and Resourcing

James R. Broyles, Paul Emslie, Bart E. Bennett, and Lisa M. Harrington

8. Introduction

The Air Force requires a highly varied composition of trained personnel, and the training of such personnel is complex and multidimensional. Air Education and Training Command (AETC) is largely responsible for providing initial skills training for airmen across a variety of officer and enlisted Air Force Specialty Codes (AFSCs). Each specialty has unique training requirements in terms of training delivery systems, the number and length of courses, the number and skills of training instructors, and the facilities required. Apart from the training content, the delivery of training in such a large and complex system requires multiyear planning, obtaining and distributing resources, adjusting to changes in training needs, and managing the movement of students from intake to graduation. In the end, the goal of the overall process is to ensure that career fields have sufficient manpower for each AFSC.

RAND's FY 2016 research suggested that technical training pipelines could benefit from analytical methods used in supply chain management because of the similarity between a private organization's supply chain and the AETC technical training process. Both systems face uncertainties and must make decisions amidst uncertainty—including uncertainties in demand, supply availability, and resourcing constraints. This section draws on best practices found in the supply chain and production literature to develop a resourcing decisionmaking framework specific to the technical training environment. Specifically, this section

- draws parallels between the Air Force technical training process and the supply chain and production systems
- summarizes approaches and best practices regarding resource decisionmaking from the supply chain and production literature
- provides a resourcing approach that enumerates all resourcing options, PGL change scenarios, and estimates the performance of each resourcing option.

The intent here is to identify best practices about resource decisions. Generally, resourcing approaches can be categorized as either proactive, where additional resources are secured beyond what is needed to meet current requirements, or reactive, which relies on swift processes to secure resources once requirement changes are observed. Such decisions are stratified based on uncertainty and risk, and the best solutions are those that maximize resilience to risk. The literature describes resiliency as the proper balance between a robust (i.e., proactive resourcing) and an agile (i.e., reactive resourcing) production system.[6] In alignment with the literature and

---

[6] The literature uses the terms *robust* and *agile* and, therefore, we use these terms when discussing the literature. However, when discussing technical training, we use the terms *proactive* and *reactive* as synonyms for *robust* and *agile*, respectively, because we feel they more naturally describe technical training resourcing decisions.

used in this document, the following are definitions used for the terms *robustness*, *agility*, and *resiliency* (Durach, Wieland, and Machuca, 2015):

- *Resource robustness*: A robust system is one that can appropriately handle unforeseen changes in demand, supply, and requirements given the current magnitude and distribution of resources. Robustness is typically obtained via *proactively* securing resource reserves beyond the expected resource requirement. Robustness is enabled by proactive resourcing.
- *Resource agility:* An agile system is one that can appropriately handle unforeseen changes in demand, supply, and requirements via flexible processes that can rapidly and dynamically respond to the environment. Agility is reactive resourcing and is typically obtained via processes that *reactively* scale or reallocate resources in response to observed changes.
- *Resource resiliency:* A resilient system is one that balances robust and agile resourcing approaches given estimates of uncertainty and risk relevant to each resourcing decision. Resilient decisionmaking is typically obtained by considering robustness and agility constraints of the resources and resourcing processes.

In the next section, we discuss how the nonrated technical training pipeline can be envisioned as a supply chain.

9. Parallels Between the Nonrated Technical Training Process and Supply Chain and Production Systems

In general, supply chains are characterized by suppliers providing products to organizations that add value to the products and then sell to their customers. Production for each organization in the supply chain is based on information (frequently uncertain information) about the demand for their products. Supply chain management strives to integrate information among plants that perform value-added process steps (production), distribution centers (distribution), and markets (i.e., consumers) (Shapiro, 2001). Prior to modern supply chain management, supply chain organizations (i.e., suppliers, plants, distribution centers) acted independently of one another without sharing information on customer demand, inventories levels, or material availability, which caused significant uncertainty. Integrating information and encouraging cooperation across supply chain echelons enables better planning and less uncertainty in material availability and customer demand.

Within a supply chain, organizations must choose how they are going to produce their products and determine a resourcing approach that enables their chosen production system. There are several types of production systems, and the ones most relevant to technical training are (definitions adapted from Molina, Velandia, and Galeano, 2007)

- Make to order (MTO): Production is scheduled according to *observed* customer orders where their *resourcing approach is to reactively* produce products.
- Make to stock (MTS): Production is scheduled according to *forecasted* customer demand, and customer demand is met from finished product safety stock. In MTS production, the *resourcing approach is to proactively* produce products prior to customer orders for the product.
- Assemble to order (ATO): Standard parts and subassemblies are acquired or manufactured according to *forecasts*, while schedules for remaining components, subassemblies, and the final assembly are not executed until detailed product specifications have been derived from *observed* customer orders.

The key difference between these production systems is their use of forecasts and resourcing approaches needed to enable each production system. MTO systems do not forecast customer demand and wait until customer orders are observed and production is reactively planned to the orders. MTS systems create production plans according to a demand forecast and proactively produce a safety stock of products ready for customer delivery once customer orders are placed. ATO systems are somewhere in between. ATO systems forecast demand for some parts and maintain a safety stock for those parts. Then, once orders are observed, the parts are promptly assembled reactively to customer order specifications.

The Air Force nonrated technical training process can be viewed as two supply chains that have different underlying production systems and resourcing approaches: a supply chain for creating instructors and a supply chain for creating trainees. At the end of each supply chain, the instructors are matched with the trainees at the schoolhouses to produce graduates. Figure 9.1 is a conceptual representation of the two supply chains.

Figure 9.1. A Conceptual Representation of the Instructor and the Trainee Supply Chains

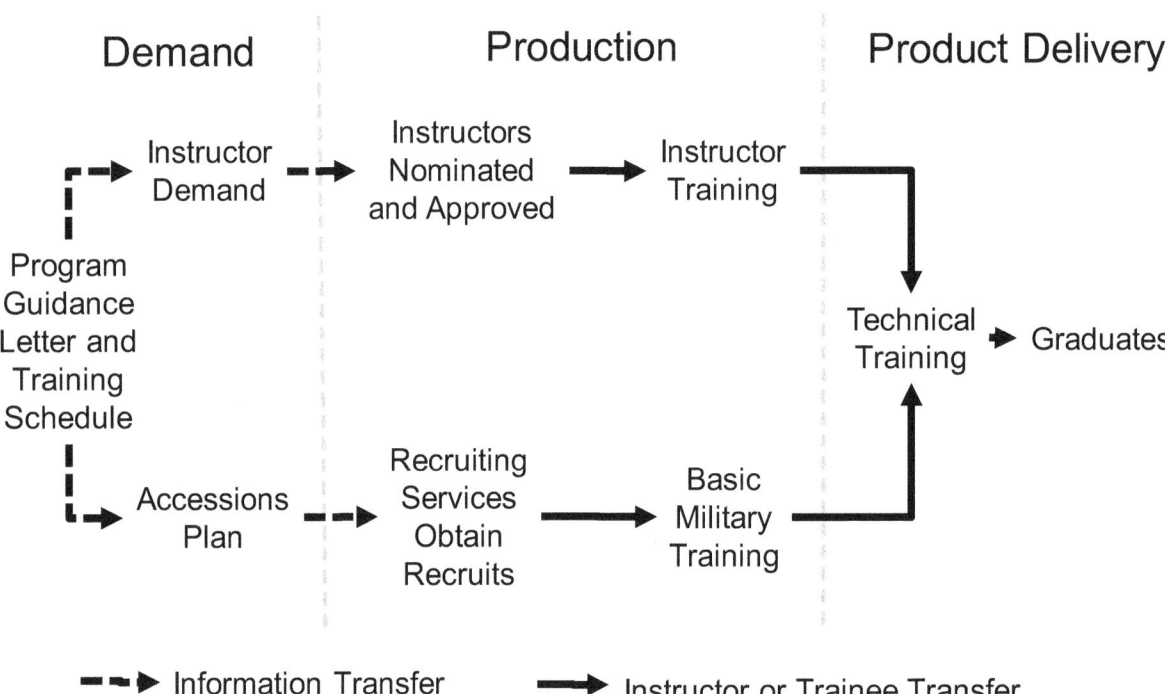

Although they must be matched to provide training, instructor production and trainee production are currently executed in the technical training process like different production systems. These differences affect the ability to react to PGL changes and will be explained in more detail in the next two sections.[7]

Instructor Production System and Reactive Resourcing

The instructor production system, represented in Figure 9.2, operates as an MTO system where instructors are acquired per the PGL guidance and training schedule and where the PGL and training schedule are treated as a customer order. This makes the instructor decisions and

---

[7] The PGL contains guidance from Air Force, Manpower, Personnel and Services (AF/A1) about specific student numbers required for each AFSC each year. This information is the demand signal for both the instructor and trainee production systems.

production a reactive resourcing approach that requires a given lead time to find and train the instructors for teaching.[8]

Figure 9.2. Current Instructor Make-to-Order Production System

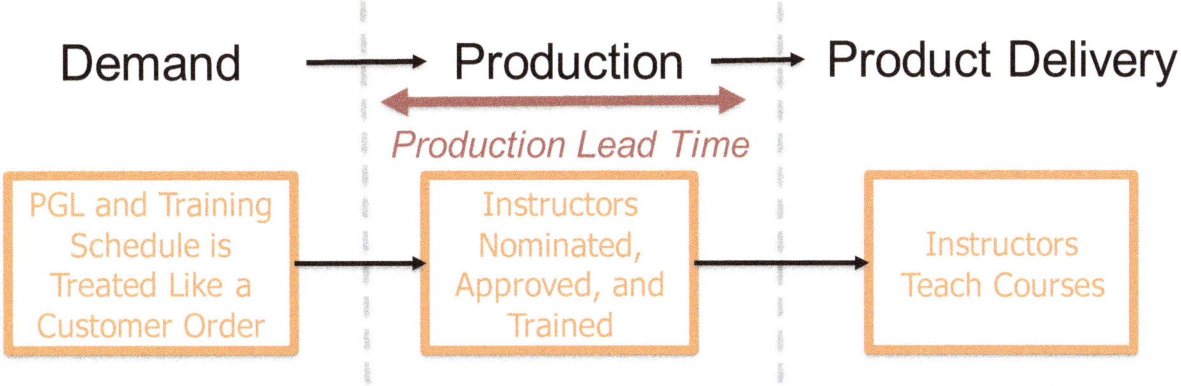

Demand → Production → Product Delivery

Production Lead Time

- PGL and Training Schedule is Treated Like a Customer Order
- Instructors Nominated, Approved, and Trained
- Instructors Teach Courses

To produce instructors, operational squadrons nominate potential instructors to the DSD process. The DSD process screens the nominees and determines eligibility and approval, the approved nominees are trained by the instructor training process, and the instructors are then delivered to training squadrons to learn the course-specific material, practice, and begin teaching. The time it takes to complete the DSD process and instructor training is represented as a "production lead time" in the figure and is the shortest time in which new instructors can be produced. In practice, this lead time often requires approximately nine months but can be longer (e.g., 18 months), depending on the specific pipeline.

During this lead time, the nominees' military record is checked to ensure they meet a performance rating and military conduct standard as part of the DSD process. The nominees then have to be given time to arrive at their duty station, go through formal instructor training, observe specific course training, and be observed successfully teaching all of their classes before being fully certified. This process may vary by AFSC, which may have specific requirements that are more or less complex, and by the individual instructor's teaching capabilities.

The PGL is determined by the Air Staff who estimate future requirements based on the information obtained from the functional managers and congressional end strength requirements one to two years in advance of its execution. As a result, the PGL is a *forecast*, rather than an exact customer order. Like any demand forecast, there is forecast uncertainty, meaning that the PGL can fluctuate for any given AFSC or as a total, and the final demand can be significantly different from the forecast. Nonetheless, although the PGL represents a forecast, specific "orders" for instructors are placed based on those forecasts. The lead time in the current process,

---

[8] Instructor assignments are typically three to five years. It is the decision to increase the number of instructors that is reactive to the PGL and to the departure schedule of the current instructors.

however, makes it impossible to increase the production of instructors on short notice, which limits AETC's ability to respond in a timely way to PGL changes.

Figure 9.3 quantifies the historical changes in the enlisted PGL. The lines in the figure are the percentage of enlisted AFSCs that have experienced at least a 5, 15, 25, and 35 percent increase in student requirements. This analysis is limited to percentage increases in this figure because understaffing instructors as the result of a PGL increase prevents AETC from attaining its training targets, which is its primary concern.[9] A PGL change is calculated as the change from the initial PGL to the most recent PGL requirements for each AFSC. For example, if the initial FY 2019 PGL for an AFSC is 120, and it increases to 150, then the percentage increase is 25 (i.e., [150 – 120]/120 = 25). Figure 9.3 shows that between 8 percent and 76 percent of the AFSCs, depending on the year, had an increase in the trained personnel requirement of at least 5 percent.

Figure 9.3. Percentage of Enlisted Air Force Specialty Codes with Increases in the Trained Personnel Requirement as Stated in the Program Guidance Letter

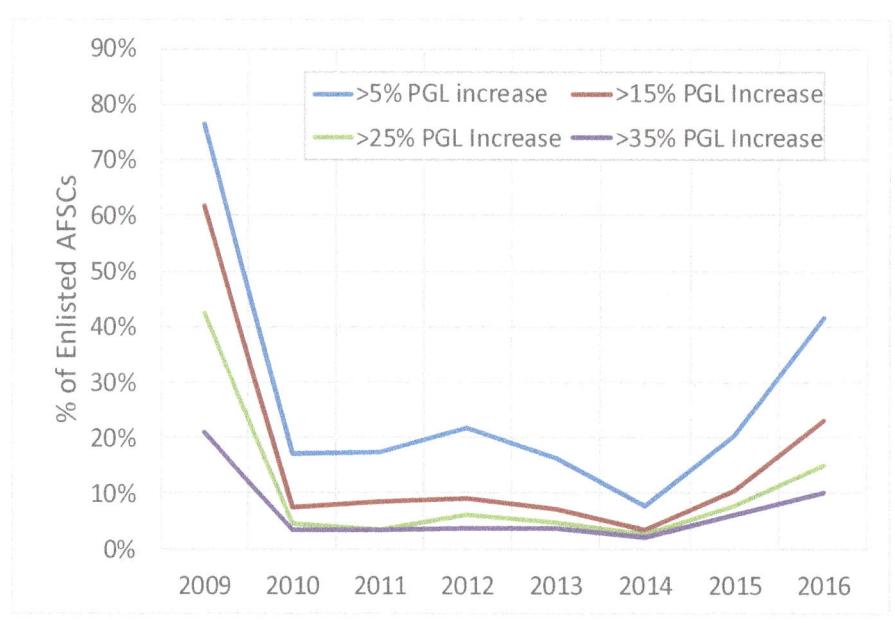

Figure 9.3 also shows a large increase in PGL changes both in magnitude and prevalence across the enlisted AFSCs in 2009 and 2016.[10] Although some schoolhouses may have available seats and be able to handle small increases in the PGL without adding instructors, the intent is to currently staff instructors as closely as possible to the number of graduates required by the PGL to avoid an excess or shortage of instructors. In 2016, 10 percent of the AFSCs had an increase in

---

[9] PGL decreases will yield overstaffing which is a misalignment of resources but will not prevent them from reaching training targets.

[10] Figure B.1 in Appendix B shows a figure similar to Figure 9.3 for officer AFSCs.

requirements for trained personnel of at least 35 percent (increases were at or below 6 percent since 2010) with 41 percent of the AFSCs having at least a 5 percent increase. The percentage of AFSCs that had a greater than 15 percent increase rose to 23 percent in 2016 from a low of 3 percent in 2014.[11]

Converting the instructor production system from an MTO to an MTS system is one way to increase flexibility to respond in a timely way to PGL changes. A resourcing approach that proactively produces instructors will enable the MTS system. An MTS system would treat the PGL as a forecast and create a safety stock of instructors that can be called upon in short order. Figure 9.4 is a representation of an MTS instructor production system. Later in this section, we provide an analysis of how a safety stock of instructors could handle requirement increases for the Intelligence Officer pipeline (AFSC 14NX, where X is the different qualification levels for the career level).

Figure 9.4. Proposed Instructor Made-to-Stock Production System

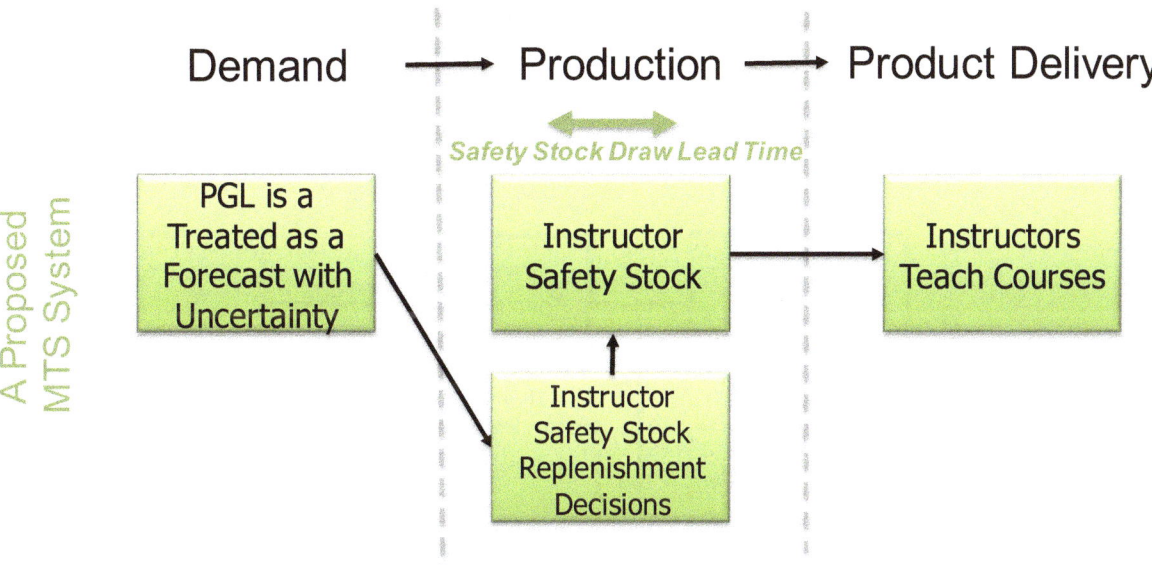

The instructor safety stock need not be at the training squadrons awaiting a potential PGL change; this would be expensive and wasteful. However, the safety stock of instructors could have another assignment and then be called up if needed as instructors. Doing so may substantially decrease the lead time it takes to get an instructor from an operational unit into a training unit as a fully qualified instructor, which would provide greater flexibility to meet changes in demand. In Part III, Section 11, we enumerate ways to obtain instructors and provide

---

[11] These perturbations were due to a sharp increase in Air Force end strength in FY 2016 that was mandated by Congress after several years of reductions in end strength (see Harrington et al., 2017, for more detail on these changes).

an introductory discussion on how this may be approached. A more detailed discussion of considerations in how to create an instructor pool is presented in Part IV.

Trainee Production System and Flexible Resourcing

The trainee production system allows more timely responses to PGL changes than the instructor production because of some flexibility in the AFSC assignment process. In contrast to instructor production, the trainee production is a mix of an MTO system for Guaranteed Training Enlistment Program (GTEP) recruits and an ATO system for non-GTEP recruits, which enables flexibility (Figure 9.5).[12] The 2AF creates an accession plan for both GTEP and non-GTEP recruits that is informed by the PGL and the training schedule. Recruiting Service obtains recruits and supplies them to BMT. BMT trains the recruits and produces airmen for the technical training squadrons. Then, instructors at the training squadrons teach and produce graduates.

Figure 9.5. Trainee Production System for Guaranteed Training Enlistment Program and Non–Guaranteed Training Enlistment Program Airmen

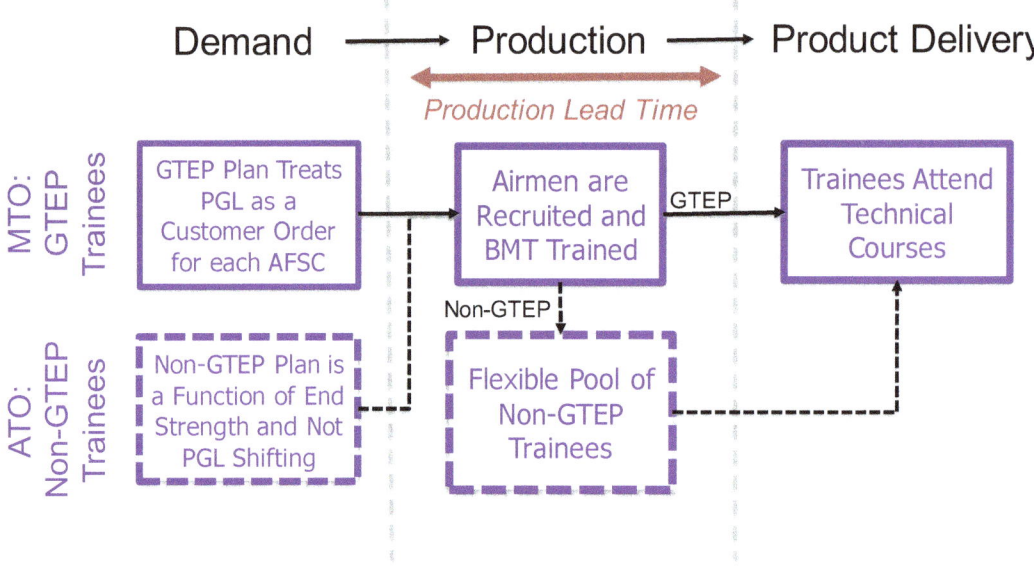

For GTEP recruiting, the PGL and the target GTEP proportions are treated as an exact customer order like in an MTO system; GTEP recruits are obtained to hit PGL established goals

---

[12] In the GTEP, specific people are matched to specific positions, and recruits are guaranteed to enter their selected AFSCs, provided they are not disqualified during training. Non-GTEP recruits are brought in under a general enlistment and can be assigned to any AFSC according to the needs of the Air Force (Air Force Instruction 36-2101, 2017). According to discussions with Air Force Recruiting Service, GTEP recruits usually account for 60–65 percent of the recruits in a year, while 35–40 percent are brought in under general recruitment.

by AFSC. However, non-GTEP production is like an ATO system, where non-GTEP airmen are produced and then assigned to (i.e., "assembled with") their AFSC during the assignment process (referred to as the job spin[13]) shortly before the start of technical training. This late AFSC assignment for non-GTEP airmen enables a timely response to PGL change within a given end strength. This flexibility is not a characteristic of the instructor production system. Therefore, the trainee production system can respond more quickly to changes in the PGL than the instructor production system. To change the instructor production system, the Air Force must decide if a reactive (i.e., MTO production) or proactive (i.e., MTS production) resourcing approach is best for each AFSC, which is discussed in the following sections.

---

[13] The "job spin" is an assignment process that matches non-GTEP trainees to AFSCs with available seats.

10. Resourcing Approaches to Supply Chain and Production Systems

According to the supply chain and production systems literature, organizations should strive for resilience in their supply chain and production systems. As described in Part III, Section 8, resilience is achieved by balancing robustness and agility (Durach, Wieland, and Machuca, 2015; Melnyk et al., 2014; Sáenz and Revilla, 2014). A robust system is one that can appropriately handle unforeseen changes in demand, supply, and product specification given the current magnitude and distribution of resources. Robustness is typically obtained via proactively creating a resource safety stock beyond the expected resource requirement (Azadegan et al., 2013), such as is done in an MTS production system (Molina, Velandia, and Galeano, 2007).

An agile system is one that can appropriately handle unforeseen changes in demand, supply, and requirements via flexible processes that can rapidly and dynamically change the magnitude and distribution of resources (Christopher and Towill, 2001). Agility is typically obtained via swift processes that reactively scale or reallocate resources and tactics in response to observed changes (Gligor and Holcomb, 2012). An MTO production system is an example where swift processes enable reasonable production lead times after a customer order is placed. This section discusses each resourcing strategy in turn and how to determine which is most appropriate.

Robustness as an Enabler to Supply Chain Resilience

Some literature describes robustness as a better approach than agility because proactive risk mitigation is more effective than reactiveness (Kleindorfer and Saad, 2005). One survey found that most supply chain managers prefer robustness to agility because they consider agility "expensive and uncertain" (Lavastre, Gunasekaran, and Spalanzani, 2012, p. 835).

The dimensions of robustness include a risk-management orientation and visibility of potential changes (Durach, Wieland, and Machuca, 2015; Christopher and Lee, 2004). Risk-management orientation involves estimating uncertainty and designating degrees of potential uncertainty to various aspects of the supply chain (Bhattacharya et al., 2013; Craighead et al., 2007), such as uncertainty in supplier performance, demand, and production capability. Visibility is key to identifying points of uncertainty in various elements of the supply chain (Christopher and Lee, 2004).

A supply chain management best practice is to share production-related information across the entire supply chain and the involved organizations. Information sharing can be done in many different ways, with varying frequencies and with varying degrees of information granularities. At one extreme, an organization may not share any information of customer demand trends with their suppliers and may not share any production lead time information with their customers. At the other extreme, an organization's information technology system communicates in real time with suppliers and customers. Once information becomes visible, managers have a better sense of supply and demand uncertainty and can better allocate resources to handle uncertainties (Lavastre, Gunasekaran, and Spalanzani, 2012). For example, if an organization is uncertain about the demand magnitude for a new product, then they may choose to add additional production resources to handle unexpected demand increases.

In the context of technical training, visibility would include visibility of the potential for requirement changes (i.e., demand changes), number of instructors available, timeliness to add new instructors, and accessions progress. Early indications of changes in requirements, resource availability, and accessions *prior* to changes being finalized can help decisionmakers choose resource strategies, suggesting real-time communications across all points of the process would be of great value.

Agility as an Enabler to Supply Chain Resilience

An agile system has swift processes to rapidly scale and distribute resources in response to an observed change. In general, agility is viewed as less desirable than robustness because dynamically scaling resources based on unpredicted changes can be highly disruptive and expensive. But, in some cases, an agile resourcing approach may be preferred to a robust one when resources are expensive to obtain and obtaining a safety stock of resources is not financially feasible. A potential example is using contractors to deliver technical training. If contingency contracts are written such that contractors must swiftly increase their instruction capacity to respond to requirement changes in short order, then the contractors may charge additional fees to enable this swift response capability.

According to the literature, the dimensions of agility generally include alertness, accessibility, decisiveness, swiftness, and flexibility (Gligor, Holcomb, and Stank, 2013). Alertness, accessibility, and decisiveness are categorized as "cognitive" dimensions in that they rely on information processing (Gligor, Holcomb, and Stank, 2013). Swiftness and flexibility are categorized as "physical" dimensions in that they require action and implementation. The following definitions are adapted from Christopher (2000) and Gligor, Holcomb, and Stank (2013):

- *Alertness* is the ability to detect changes, opportunities, and threats so that a supply chain can respond.
- *Accessibility* is the ability to access relevant data to make an informed decision after a change is detected.
- *Decisiveness* is the ability to make decisions resolutely with the proper information.
- *Swiftness* is the ability to implement decisions quickly.
- *Flexibility* is the ability to modify the range of tactics and operations.

To implement agility, all five dimensions are needed.

In general, organizations that have agility also have a "stable backbone" to their operations that can dynamically allocate resources (Aghina, De Smet, and Weerda, 2015). The backbone includes a stable organizational and governance structure, decisionmaking, and execution processes. These organizations typically have stable resources dedicated to stable customer demand streams and dynamically deploy resources to only explore potentially risky and new products or services.

In the context of technical training, *agility* means swiftly adding or acquiring instructors, training instruments, classrooms, and/or dormitory space once changes in requirements or accession plans are observed. *Alertness*, *accessibility*, and *decisiveness* as described by the literature relate to the ability to detect (or even forecast) PGL changes and have visibility of the current training capacity. *Swiftness* and *flexibility* relate to the ability to change instructor requirements and to obtain additional instructors in short order. Depending on the time frames, obtaining true agility in technical training is likely to be difficult because of the long lead times to add and redistribute resources. However, there may be certain instances where a reactive approach (agility) is better than a proactive approach (robustness).

Deciding Between a Robust and an Agile Resourcing Strategy

The level of uncertainty, resource costs, and timelines form the boundary conditions for deciding resourcing strategies (Chopra, Reinhardt, and Mohan, 2007). For example, a proactive, robust resourcing strategy is likely the best in a situation with long resource acquisition durations and highly uncertain environments (e.g., uncertain demand and supply). By contrast, an agile approach is likely the best in situations with short resource acquisition durations and generally more stable environments. Extrapolating from the literature, there are several categories that must be visible and must be considered when determining feasibility bounds and selecting the best resourcing approach:

- The magnitude of available resources and the feasibility of obtaining additional resources in a timely way. Given all resources required, determine the binding resource that limits production at each point in time given the timeliness of obtaining additional resources.
- The feasibility comparison between the resource acquisition timeliness and acceptable lead times to respond to changes in demand or supply. There are situations where acceptable lead times to respond to observed changes are shorter than the acquisition time for additional resources.
- Within the realm of feasibility, the cost to acquire resources as a function of magnitude and timeliness should be considered. It may be feasible to add additional resources, but the costs—including financial and workload costs—for doing so may be very large.

The decision for AETC is to determine a resourcing approach for each AFSC. To do this, it needs to decide for which AFSCs a reactive resourcing approach (i.e., an agile approach) is sufficient and for which AFSCs a proactive resourcing approach (i.e., a robust approach) is sufficient. The choice is dependent on AETC's desire to react to requirement changes and its ability to obtain resources in a timely manner. The next section proposes a process for determining resourcing strategies.

11. A Proposed Process for Choosing Resourcing Strategies

This section proposes an approach to select a resourcing strategy that is best for each AFSC and estimates the effects of implementing the resourcing strategy being considered. With a defined resourcing strategy, decisionmakers in the technical training process will be able to understand their resource constraints and estimate production outcomes given alternate resourcing decisions. Although AETC has implemented some elements of this process, as will be described throughout this section, our discussions with stakeholders suggest that they still lack a systematic and comprehensive process that is informed by best practices. We propose a five-step process:

1. Categorize AFSCs: Determine AFSCs (technical training pipeline for a given AFSC) for which it is desired to at least consider a proactive resourcing strategy to enable quicker responses to PGL changes and AFSCs where a completely reactive strategy is sufficient given current lead times to produce or obtain more resources/instructors.
2. Quantify resourcing bounds: For AFSCs where a proactive strategy is being considered, determine resource bounds and costs by quantifying the current resources available and the magnitude and timeliness for adding resources.
3. Estimate probability for requirement changes: Determine if future requirement changes and/or accession plan changes are being considered or can be predicted.
4. Estimate effects of potential resourcing strategies: Using modeling, estimate the caseload and wait time effects of implementing considered resourcing strategies given resource bounds (from Step 2) and the potential for future required changes (from Step 3).
5. Choose a resourcing strategy: Choose a resourcing approach given the resource bounds, potential for requirement changes, and the estimated effects on caseload and wait time.

Although the steps are presented as sequential, in practice, these steps may be iterative, as described in the following sections.

Step 1: Categorize AFSCs

In Step 1, AFSCs are categorized to identify those for which a completely reactive approach to resourcing is acceptable and those for which it is important to consider proactive resourcing strategies. We recommend two categories: (1) reactive resourcing is adequate and (2) proactive resourcing should be considered. A completely reactive approach will, by definition, always lag behind the requirements signal if the requirements change before the resources can be added to meet the requirements.

For some AFCS, a reactive resourcing strategy is acceptable because of the types of skills produced and how critical the pipelines are to supporting the Air Force's major near-term objectives. For example, a reactive resourcing strategy might be appropriate for Security Forces (AFSC 3P0X1) because it is a very large schoolhouse and likely could absorb additional entries by adding seats. Furthermore, there are many things that can be done at the base level to deal with a temporary shortage of security forces. For other pipelines, AETC should consider a proactive resourcing strategy. An example here might be Tactical Aircraft Maintenance (AFSC 2A3X3) because these positions contribute to sortie generation, and a shortage would become evident at some point. Moreover, the method of instruction includes small groups working on simulated/actual aircraft equipment, so it may be difficult to increase graduates without considerable lead time. To help determine the appropriate resourcing strategy, one must answer the question: If requirements or accessions plans were to change for a particular pipeline, how critical is it to respond in a timely way in order to meet critical Air Force mission objectives?

Step 1 should be completed in parallel and iteratively with Step 2, which quantifies the feasibility, magnitude, and timeliness of adding resources. If it is only acceptable to respond to requirement changes within X months but feasible resourcing bounds require more than X months, then the pipeline should be considered for a proactive resourcing approach. If, however, resources can be increased sufficiently within X months, then perhaps a reactive resourcing strategy is adequately responsive.

Step 2: Quantify Resourcing Bounds

Step 2 quantifies the production capacity of the current resources and the bounds for adding resources in the future. Bounding future resource capacity has two dimensions: magnitude of the resources and timeliness. The magnitude of resource bounds is the maximum number of resources by resource type that can be feasibly added to the technical training process. In other words, how many instructors can be added for a particular course, given the DSD process for military instructors, or by using contractors, civilians, or other means? How many pieces of training equipment can be added? And how many additional classrooms and dormitories can be added? Of course, the answers to these questions depend on the time frame being considered. Thus, the dimension of timeliness is equally important and should be estimated.

Gathering this information typically requires data calls and/or estimates from the schoolhouses.[14] Ideally, this information would be collected, updated, and stored in a centralized database where Headquarters AETC, the 2AF, and individual training wings could look at the feasible magnitude and timeliness of additional resources by resourcing type.

Once this information is collected and centralized, it is possible to identify the "binding resource" at all points in time. To do so, the schoolhouses need to estimate the maximum number of students that each resource type (e.g., instructors, training devices, classroom size) can serve, without consideration of constraints for other resource types. The binding constraint at a particular point in time is the resource with the minimum capacity.

---

[14] One way that this is currently done is with a constraint worksheet where the schools are tasked to obtain information on their training resource capacity.

Notional Example of Step 2

Figure 11.1 shows a notional example of resource capacity and the binding resource for a hypothetical example with two resource types: instructors (squares) and training devices (plus signs). In the figure, the binding resource capacity (red line) is the minimum of the two resource capacities, which can change for different periods (as indicated on the x axis). For example, instructors are the binding resource during time frame B, but training devices are the binding resource in time frame C.

Figure 11.1. Notional Example of Binding Resource Capacity

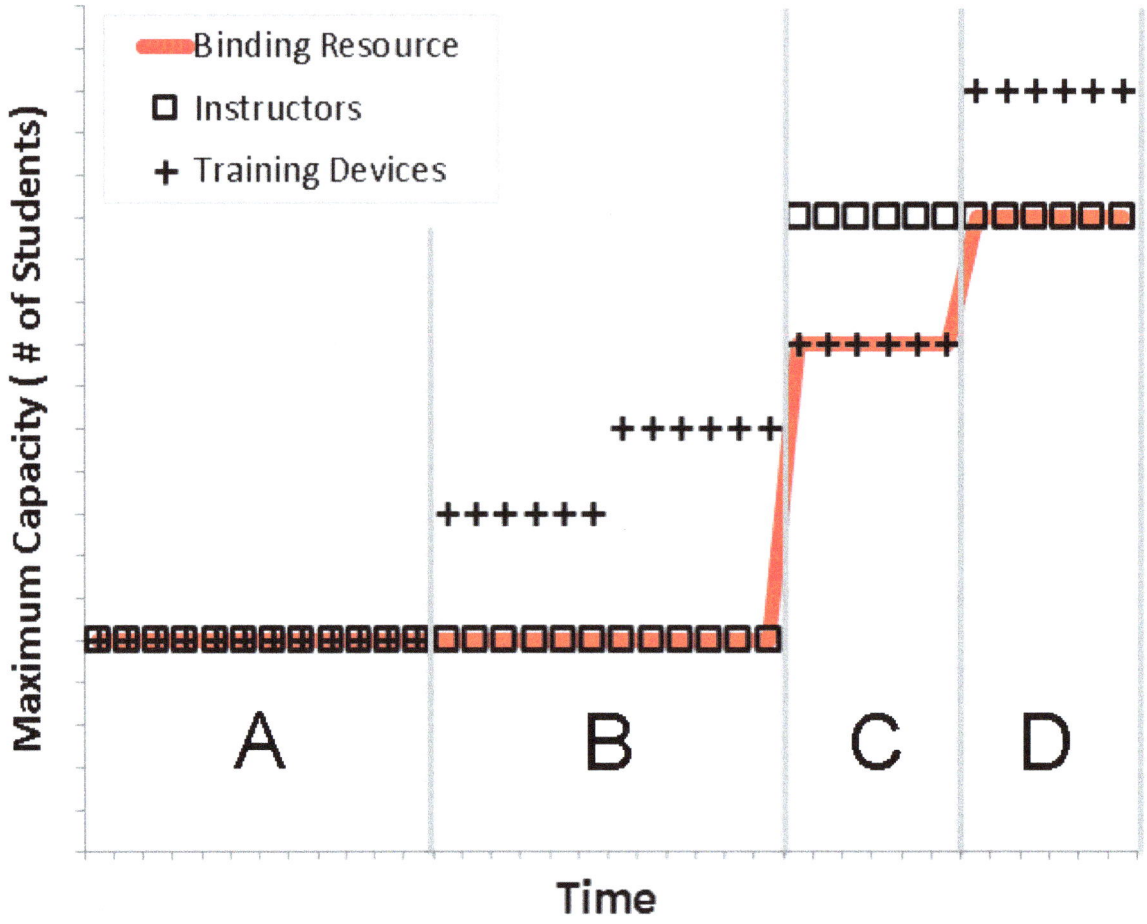

In this notional example, it is possible to add resources, but the timelines at which the resources become available differ. In time period *A*, the production lead time for both resources are greater than the length of time period A, so no additional resources can be added. In time period *B*, it is feasible to add training devices, as shown by the increase in capacity for training devices at the beginning of period B and in the middle of period B; but the number of instructors cannot be increased. Therefore, the instructors are the binding resource. In time period *C*, it is feasible to add instructors and training devices. However, more instructors can be added than the training devices can accommodate. Therefore, the training devices are the binding resource in period C.

The binding resource may need to be evaluated for each course when there are a series of courses within a pipeline. Different courses within the series may require different instructor types and training devices. In the notional example in Figure 11.1, one could add marginal capacities for each resource and course (e.g., instructors for course 1, instructors for course 2). In this case, the binding resource would be the minimum across the resource types and across the courses in the series.

The next consideration is whether the binding constraint capacity meets the timeliness requirements for future changes in requirements and accession plans. Step 4 answers this question. In Step 4, the binding resource capacity is compared with timeliness requirements (or even potential timeliness requirements) for cumulative production as specified by Air Force leadership. In short, if the binding resource capacity is sufficient to meet timeliness requirements, then a reactive resource strategy is feasible. If the binding resource capacity is not sufficient, then a proactive resource strategy is likely the prudent approach.

The binding resource capacity profile would help AETC identify which resources could be added most quickly to increase pipeline capacity, thereby increasing agility in the process. For example, if the process duration for adding instructors can be reduced such that instructors could have been added at the beginning of period B in Figure 11.1, then that would improve the timeliness.

Options for Increasing Instructor Availability

Step 2 assesses resource availability and timeliness for adding resources. Increasing available resources proactively could be a desirable approach if the assessed magnitude and timeliness of adding resources is not sufficient or desirable as is. We discuss options for proactively adding instructors and potential benefits and drawbacks for each, informed by our discussions with AETC, the 2AF, and several training squadrons. Options to proactively increase instructor availability include the following (see Part IV for more on instructor options, their benefits and drawbacks):

1. Enhance the efficiency of the DSD process.
2. Call on former technical training instructors.
3. Use contract instructors or contracting out training entirely.
4. Use civilian instructors.
5. Employ virtual and alternative methods to deliver course content.

The DSD process can be used to increase instructor availability in a number of ways: proactively identify and process instructors through the DSD process beyond what is required to meet the current training schedule, improve the DSD process to yield more eligible instructors, and/or decrease the DSD process duration. The Air Force has made progress in this area.

The DSD process is approximately six to nine months in duration, in which exemplary airmen are nominated to be technical training instructors. In June 2017, AETC announced changes to the DSD process where eligibility is open to staff sergeants and physical training score requirements are lowered from 80 to 75 (Holliday, 2017). These changes naturally increase the available pool of instructors and could thereby increase pipeline capacity in situations where instructor availability is a binding constraint.

Reducing the DSD process duration could make a reactive approach feasible for more pipelines and make pipelines already taking that approach even more agile. Similarly, processing more applicants than required would effectively shorten wait time when a requirement increases, since they will already be part way through the pipeline. Some applicants will have wasted time and effort being approved for an instructor assignment that they will not ultimately be assigned to. Because instructors are only qualified to teach their own AFSC, they are not fungible and cannot be reassigned to instruct another AFSC. This lack of fungibility makes the production of extra instructors less attractive because the excess instructors cannot be reassigned to instruct another AFSC's course.

AETC is starting to be more proactive with their instructors by starting the DSD nomination process before PGL requirements are determined and by increasing the number of instructor requirements for FY 2019 beyond that of the anticipated instructor requirements. This is applied only to a select set of AFSCs, however, based on priority of that career field. Doing so is a proactive resourcing approach given that the FY 2019 instructor requirements are undetermined.

Calling on prior instructors is advantageous because they have already been through the DSD process and instructor training. Training squadrons that we talked with like this option and at least one squadron is trying to implement this capability. The disadvantage is that prior instructors may not want to return to instructing after a previous three-year tour, and this was a concern expressed by the training squadrons. Moreover, there is concern and perception that an instructor assignment does not advance a person's career or promotion appeal.

Contracts have the potential to significantly increase instructor availability and increase timeliness provided that contract costs are reasonable and acquiring contracts is feasible and desirable. Contracts can be created in many forms and can be broadly categorized as either contracts for instructor labor only or contracts that are responsible for all instruction delivered to meet a learning standard set by the Air Force. To increase agility, contracts (whether contracting out instruction entirely or contracting instructor labor only) may include requirements regarding response timeliness to requirement changes. If contracting labor only, an indefinite delivery/indefinite quantity contract may be a mechanism to rapidly flex instructor labor in response to requirement changes.[15] The drawbacks to using contracts (whether it is labor only or

---

[15] Indefinite delivery/indefinite quantity contract types do not specify the amount of goods or services to be demanded over a finite period of time. These contract types allow the government to request goods and services as needed.

otherwise) is that they may be too expensive, they may not be able to attract necessary labor, and the Air Force may want to maintain organic instruction capabilities because of the nature of the content. Depending on the location of the schoolhouse, it may be difficult to attract contract labor to locations where civilians may not want to live. In addition, the skills required to teach some AFSC courses may also be desirable in civilian jobs. This is particularly true for cyber AFSCs—private employers seek to hire in this field.

Hiring civilian instructors is another option, but it suffers the same drawbacks as contractors. In addition, the relatively low wages earned by uniformed instructors is frequently not sufficient to attract a civilian who may be able to command better pay in private employment, depending on the career field.

Virtual and alternative methods to deliver content offer the potential for more efficient use of existing instructors, reducing instructor requirements, and increasing the pool of available instructors. Alternative methods include guided self-study and examinations and the ability to test out of and into blocks and courses. Virtual instruction naturally reduces drawbacks of military or civilian instructors needing to move to the schoolhouse. It also may reduce the requirement to move students to the schoolhouses (see Part II for more detail on alternative content delivery options).

Step 3: Determine Potential for Requirement Changes

Step 3 determines potential for changes in the accession plan and the PGL. This step includes changes that have been finalized already and changes that are being considered by Air Force leadership. Having visibility of potential changes in combination with the AFSC priorities in Step 1 can help AETC select appropriate resourcing strategies.

We propose two methods to obtain indications of potential requirement changes: qualitatively, based on early indication about changes from Headquarters Air Force, and quantitatively, in which requirement changes may be indicated from historical PGL data and/or from planned end strength changes. For the qualitative approach, three pieces of potential change information are needed: magnitude of the change, acceptable response timeliness, and likelihood that the change will occur. For example, assume that a particular AFSC is of high enough priority for AETC to take a proactive resourcing approach. Then potential change information could indicate it is highly likely that an X-percent requirement increase will be finalized and will need to be addressed within Y months.

For the quantitative approach, we propose and show data analysis designed to answer the following two questions:

> Q1: On average, which AFSCs have decreases, increases, or approximately no change in their training requirements from initial to final PGL?
>
> Q2: Which AFSCs have highly variable and which have stable training requirements when comparing the initial with the final PGL?

This analysis is designed to identify AFSCs that have historically variable PGL changes. Using PGL data over a seven-year period, we categorize the AFSCs into the following categories:

1. Average PGL change is within ±20 percent and less than 20 percent of the years had a greater than 25 percent increase or decrease.
2. Average PGL change is within ±20 percent, and more than 20 percent of the years had a greater than 25 percent increase or decrease.
3. Average PGL change is greater than a 20 percent increase.
4. Average PGL change is greater than a 20 percent decrease.

Category 1 AFSCs are those that have relatively stable PGLs. Category 2 AFSCs are historically uncertain and have had PGL changes that show both large increases and large decreases. Category 3 and 4 AFSCs have historically uncertain PGL changes that tend to have large increases or decreases, respectively.

Table 11.1 shows the AFSCs grouped in the four categories using PGL change data from 2009 to 2017.[16] Table 11.3 includes the 189 AFSCs that had at least four years of populated data with an initial PGL.[17] Our assessment shows that 138 (73 percent) of them are relatively stable, falling into Category 1; 33 (17 percent) have a history of uncertain requirements and fall into Category 2.

Table 11.1. Number of AFSCs in Each PGL Change Category

Average PGL Change	Enlisted	Officer	Total
Within 20% and less than 20% of the years had a greater than 25% increase or decrease	122	16	138
Within ±20% and more than 20% of the years had a greater than 25% increase or decrease	25	8	33
Greater than 20% increase	5	0	5
Greater than 20% decrease	11	2	13
Total	163	26	189

AFSC 1N331 Cryptologic Language Analyst, which falls in Category 2, provides an interesting example, as shown in Table 11.2. From 2010 to 2012, 1N331 had large increases in PGL requirements and, from 2013 to 2016, had large decreases averaging out to about a 10 percent change. In addition, three of the seven years (43 percent) had a greater than 25 percent

---

[16] We obtained enlisted PGL data from 2009 to 2016 only where 2017 was omitted. We obtained officer PGL data from 2010 to 2017 where 2014 was omitted.

[17] Some AFSCs had a zero initial PGL, and a PGL change increased the PGL to a positive number. These cases were omitted, and only cases where the initial PGL was greater than zero were included in the analysis. Of each year-AFSC combination from 2010 through 2017, about 7 percent had an initial PGL value of 0 and had a PGL change subsequently.

increase or decrease. 1N331 is an AFSC that is historically unpredictable and, if the future is similar to the past, there will significant uncertainty in requirements. But whether that uncertainty will look like the recent past is in question.

Table 11.2. Requirements Changes for Cryptologic Language Analyst

AFSC	Initial PGL (PGL Change Quantity, Percentage) by Year							Average Initial PGL	Average % Change	% Years of >%25 Change
	2010	2011	2012	2013	2014	2015	2016			
1N331	1,299 (106, 8%)	1374 (100, 7%)	1452 (337, 23%)	1936 (–24, –1%)	338 (–137, –41%)	258 (–138, –53%)	112 (–60, –54%)	967	–10%	29%

Summarizing the quantitative and qualitative analysis, Table 11.3 shows a notional example of requirement change scenarios, their estimated magnitude, acceptable response timeliness if requirement change were to occur, and the estimated probability of the change being observed.

Table 11.3. Notional Example of Requirement Change Scenarios

AFSC	Timeframe	Estimated Probability of Requirement Change (% Change Given Change is Realized)		Acceptable Response Timeliness Target (months)
		Qualitative Sources	Quantitative Sources	
A	Within Execution Year	Low (+10%)	Low (±10%)	3
	Prior to Execution Year	Low (+10%)	Low (±10%)	6
	Next Year	No Change Considered	High (±13%)	12
B	Within Execution Year	No Change Considered	Low (±5%)	2
	Prior to Execution Year	High (+30%)	Low (±5%)	9
	Next Year	High (-35%)	High (±33%)	12
C	Within Execution Year	No Change Considered	Low (±25%)	2
	Prior to Execution Year	No Change Considered	Low (±10%)	6
	Next Year	No Change Considered	Low (±10%)	12

For example, in Table 11.3, the qualitative sources for AFSC A, Within Execution Year, indicate a low probability that a change will happen within the execution year. But, if it does happen, the estimate is that it will be ±10 percent of the current PGL, with an acceptable training response time of three months. For AFSC B, Next Year, both qualitative and quantitative sources indicate a high probability of requirement change. If that change occurs, the quantitative sources estimate the change to be ±33 percent, and the qualitative sources estimate the change to be –35 percent, with a 12-month acceptable response target.

The "Acceptable Response Timeliness Target" column is a category that needs further definition. Does "acceptable response" mean graduating X percent of the requirement increase or starting training for *Y* percent of the requirement increase? Defining the acceptable timeliness target for the "acceptable response" is also needed.

If this potential change is well within the resourcing bounds as defined by Step 2 (both in terms of magnitude and timeliness), then a reactive resourcing approach is probably adequate. If, however, the potential change is outside of the resourcing bounds, then a proactive resourcing approach is needed to ensure timely response requirement changes.

These methods rely on a requirement range for planning purposes. Informed by both qualitative and quantitative sources, one could set the planning requirement (as done with the PGL) and then set a contingency requirement where instructing resources are poised to be secured rapidly if the additional contingency requirement is realized. This could be the completion of the DSD process for additional instructors or other methods that would create a pool of available instructors to draw on as needed.

These early indications of potential requirement changes will be inputs to Step 4.

Step 4: Estimate Effects of Potential Resourcing Strategies

Step 4 uses modeling to estimate the effects of potential resourcing strategies given hypothetical requirement change scenarios. This step takes information gathered from Steps 1, 2, and 3 as inputs, uses information on classes (e.g., durations, sequences, schedules), and estimates the production effects. We recommend estimating several production outcomes:[18]

- students in the pipeline
- students awaiting training
- average student wait time and average total production time
- cumulative graduation production.

Figure 11.2 shows a notional example of outputs from a model that estimates the number of students in the pipeline given a notional increase in accessions/requirements and different resourcing approaches. In the reactive approach, the number of training resources is increased after the requirements change is observed (red line). Because it takes time to add resources, students are assigned to classes further in the future. This is because the additional training resources are not ready immediately and take some time to develop and acquire. Therefore, more students are awaiting training initially. As the resources are added, the number of students in the pipeline will decrease. In the proactive approach, extra resources (e.g., extra instructors) are available prior to observing the requirement change (dotted blue line). The temporary increase in students in the pipeline is because of the spike in accessions. With proactive resourcing, resources are immediately available, and there are fewer students awaiting training compared with the reactive approach.

---

[18] If interested, other production outcomes could be estimated by a model such as the percentiles of wait time and production time.

Future development of a modeling approach to produce such estimates could be based on stock and flow models. These types of models estimate the number of students in the pipeline (i.e., the stock) and rates at which students enter, leave, and proceed through the system (i.e., the flow). As inputs, the models will need resourcing magnitude and timeliness (from Step 2), requirement change scenarios to model (from Step 3), and information on courses to include course durations, sequences, and schedule.

Figure 11.2. Notional Example of Students in the Pipeline Given Accession or Requirement Changes and Different Resourcing Approaches

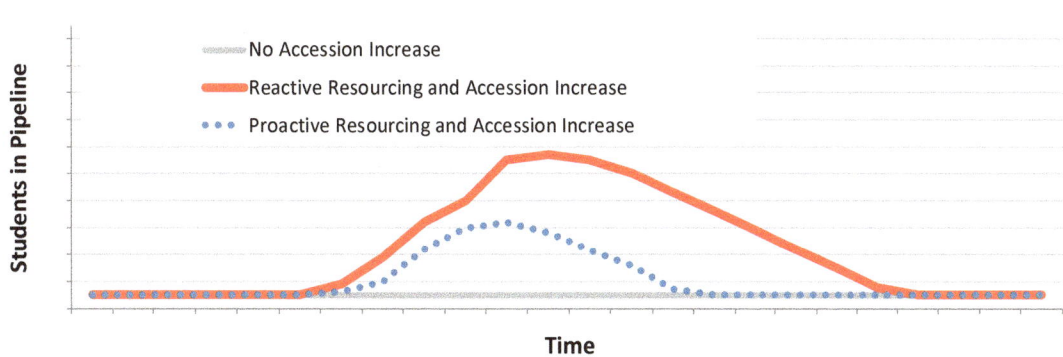

Step 5. Choose a Resourcing Strategy

In Step 5, a resourcing strategy is chosen given the information obtained in Steps 1–4. Step 5 should also consider external factors such as costs, desirability to add resources, long-term plans for particular AFSCs, and benefits and drawbacks of acquiring resources, such as instructors, through different means (such as those discussed in Part IV).

This decision environment is multidimensional and contains uncertainty, which makes it difficult to make resourcing decisions. It is multidimensional in that the decisionmaker must consider feasibility, costs, timeliness, and desirability of different resourcing strategies. It contains uncertainty because there is uncertainty in requirement changes, accessions, washbacks,[19] and eliminations. In such environments, the decisionmaker needs to assign value to the specific resourcing approaches given the multidimensional trade-offs and uncertainty.

It is possible that some resourcing strategies outperform other strategies in all dimensions. For example, consider a situation where proactive resourcing is approximately as costly as reactive resourcing, and there are no major drawbacks to pursuing proactive resourcing. In this case, proactive resourcing is the dominant solution. It is more likely that no one strategy

---

[19] *Washbacks* are students to do not pass blocks of a course and must be reinserted in to a subsequent course to repeat the failed instructional block. For example, the Intelligence Officer Course (AFSC 14N1) has a six-month course that start every two weeks. If a student fails to pass a one-week long instructional block, they can be washed back to the course that started two weeks later if there is an open spot.

dominates in all dimensions. For example, consider a situation where proactive resourcing is more costly than reactive resourcing: Proactive resourcing will respond in a much more timely way to potential requirement changes. In such a case, the decisionmaker must decide if responding more quickly to potential requirement changes is worth the additional cost. Costs include financial costs to obtain instructors in short order (e.g., contract cost), the cost in terms of person-hours required to process extra nominees through the instructor DSD process, and the readiness cost of not meeting training targets if resourcing is strictly reactionary to PGL changes.

Example of the Proposed Resourcing Process: Intelligence Officer Course (AFSC 14N1)

This section presents an example of the five-step process as applied to the Intelligence Officer Course (AFSC 14N1) and assumes that 14N1 is an AFSC for which AETC wants to consider for proactive resourcing in Step 1. The Intelligence Officer Course is the only course in this series, is six months in duration, and comprises several blocks.[20] If students do not pass a block, they are given a second chance to retest. If they do not pass their second chance, they can be washed back to another course that is approaching the block of material.

Table 11.4 shows the FY 2017 schedule of classes for 14N1 as of May 12, 2017. There is a total of 24 classes for FY 2017, and classes typically start every two weeks, although sometimes there are three weeks between class starts. "Baseline Programmed Entries" are those planned for in the original PGL. "Total Entries" is the total number of students that started the class, including first-time students and washback students. The "Observed Number of Graduates" is the total number of students that passed the course.

---

[20] This course is delivered by the 315th Training Squadron at Goodfellow Air Force Base in San Angelo, Texas.

Table 11.4. Fiscal Year 2017 Intelligence Officer (AFSC 14N1) Classes, as of May 12, 2017

Class Number	Start Date	End Date	Baseline Programmed Entries According to the PGL	Observed Total Entries	Observed Number of Graduates
2017001	10/12/2016	4/25/2017	34	44	33
2017002	10/26/2016	5/9/2017	17	17	15
2017003	11/9/2016	5/23/2017	34	26	TBD
2017004	11/30/2016	6/12/2017	17	54	TBD
2017005	1/4/2017	7/10/2017	30	23	TBD
2017006	1/18/2017	7/21/2017	17	44	TBD
2017007	2/1/2017	8/4/2017	17	22	TBD
2017008	2/15/2017	8/18/2017	17	21	TBD
2017009	3/1/2017	8/31/2017	17	15	TBD
2017010	3/15/2017	9/15/2017	17	18	TBD
2017011	3/29/2017	9/29/2017	17	22	TBD
2017012	4/12/2017	10/16/2017	17	18	TBD
2017013	4/26/2017	10/30/2017	17	18	TBD
2017014	5/10/2017	11/14/2017	17	36	TBD
2017015	5/24/2017	11/29/2017	17	TBD	TBD
2017016	6/7/2017	12/12/2017	17	TBD	TBD
2017017	6/21/2017	1/3/2018	17	TBD	TBD
2017018	7/5/2017	1/17/2018	17	TBD	TBD
2017019	7/19/2017	1/31/2018	17	TBD	TBD
2017020	8/2/2017	2/14/2018	17	TBD	TBD
2017021	8/16/2017	3/1/2018	34	TBD	TBD
2017022	8/30/2017	3/15/2018	17	TBD	TBD
2017023	9/13/2017	3/28/2018	34	TBD	TBD
2017024	9/27/2017	4/11/2018	17	TBD	TBD

A *single* 14N1 class typically has 17 to 18 students (i.e., 17 to 18 total entries). For example, class number 2017002 was a single class. Some of the classes are *double* in that they have two sessions (with two different instructors) totaling approximately 34 to 36 total entries (e.g., class number 2017014). Some classes were planned to be a single or double class according to the planned schedule but became a triple class totaling approximately 51 to 54. Class number 2017004 was scheduled to be a single class with 17 students (baseline programmed entries) but became a triple class with 54 students (total entries).

Step 2 is identifying resourcing bounds and timeliness to secure more resources. In this context, the Intelligence Officer Course uses instructors, classrooms, and computers as their resources. When we spoke with the 315th Training Squadron, they stated that instructors are almost always their binding resource. In practice, one would determine various ways to increase the number of instructors and quantify the feasible timelines of adding more, as discussed previously.[21]

For this example, we assume the following three resourcing options are feasible and being considered by AETC:

1. Status quo: Capacity remains as planned.
2. Every class is at least a double class: Every class is at least a double class while maintaining triple classes if already scheduled.
3. Double classes after March 2017: The class schedule remains as-is for classes that start in March 2017 or before. After March, all classes are at least a double class while maintaining triple classes if already scheduled.

We evaluate the performance of resourcing options relative to the following two change scenarios:

- No FY 2017 PGL change
- FY 2017 PGL increase of 10 percent.

A 10 percent increase was selected because 14N1 had a 7 percent PGL increase in FY 2016 (476 initial PGL with an increase of 34); we rounded the 7 percent up to 10 percent for illustrative purposes. In 2017, the 14N1 initial PGL was 525, with a decrease of 25 for a final PGL of 500. In practice, AETC should estimate the likelihood of each change scenario (as requested by Step 3), and these likelihood estimates would inform their resourcing option choice.

Step 4 calls for modeling such change scenarios and resourcing options to estimate the performance in each combination. In the 14N1 example, we evaluate the performance by estimating the point in time by which each resourcing option produces enough graduates to surpass the cumulative graduation requirements in each scenario.[22] For example, we estimate at which point in time every class is at least a double class, assuming there are enough trainees to produce enough graduates to surpass the FY 2017 PGL increase of 10 percent cumulative graduation requirements. With such an estimate and the costs of obtaining additional resources, AETC can make a more informed resourcing choice.

Figure 11.3 shows the estimated maximum cumulative number of graduates for the FY 2017 classes under the different resourcing options.[23] In cases where the number of graduates has not been observed yet, as displayed in Table 11.4, we estimate the number of graduates using the

---

[21] Related to identifying resourcing bounds is identifying recruiting bounds. If Recruiting Command cannot recruit additional airmen to train, then securing additional instructors is not necessary. Therefore, it is important to quantify the feasible magnitude and timeliness trajectories of recruiting additional airmen. If recruitment is not a binding constraint and Recruiting Command will be able to recruit more airmen if the requirements increase, then the resources are the binding constraint. For this example, we assume that recruitment is not a binding constraint.

[22] The cumulative number of graduates is not the only performance metric that could be estimated. Other metrics include number of students awaiting training and the cumulative number of person-hours spent awaiting training.

[23] This is not designed to be an actual forecast of the number of graduates, but it is an example of how different resourcing options perform under PGL change scenarios.

observed total entries, the baseline programmed entries, and FY 2016 estimates of washback and graduation rates. In the figure, "Status Quo" is estimated to hit "No FY 2017 PGL Change" cumulative graduates at the beginning of April 2018 (point A). This is on target because final FY 2017 class starts in September 2017 and ends in late March 2018. "Every Class is at least a Double Class" and "Double Class After March 2017" are estimated to hit that same mark in December 2017 (point B) and February 2018 (point C), respectively. However, if the 10 percent increase scenario is observed, then "Status Quo" will take an additional two months to surpass the number of graduates required (point D).

As described in Step 5, AETC can see the outcomes of different resourcing options and decide which resource options are acceptable in terms of cumulative graduates and timeliness. Other external factors should be considered in the decision, including cost of obtaining additional resources and desirability of obtaining additional resources.

Figure 11.3. Example of Maximum Cumulative Number of Graduates from Fiscal Year 2017 Intelligence Officer (AFSC 14N1) Class, by Class End Date Given Course of Action Selected

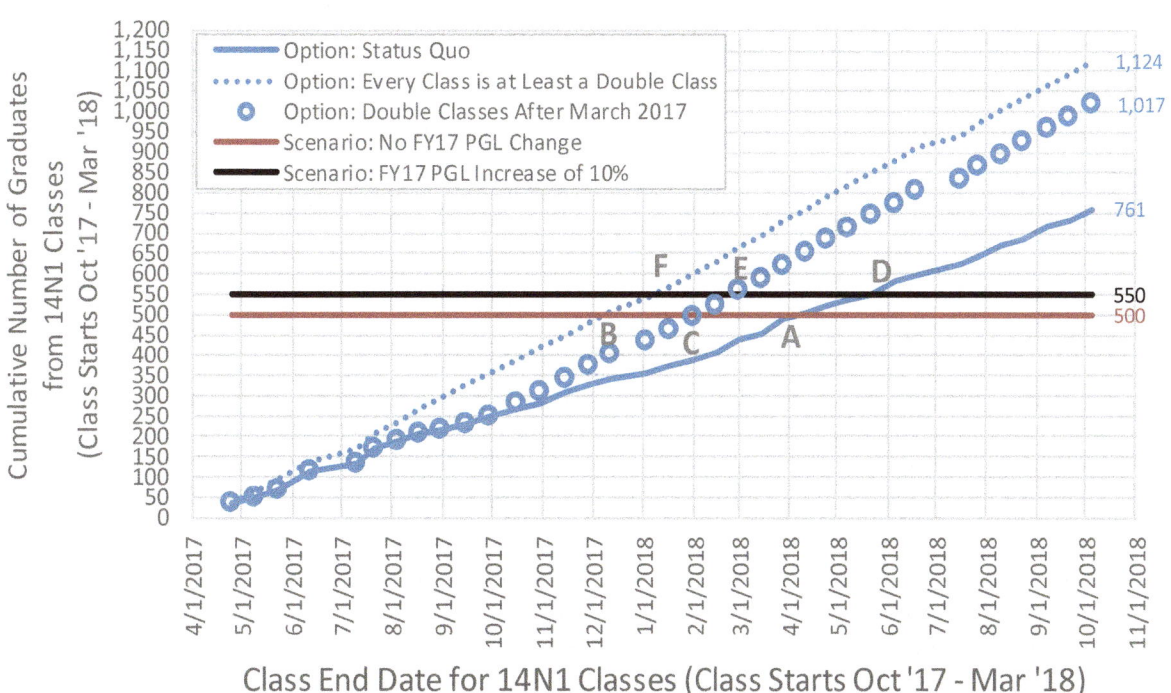

12. Summary and Recommendations

This section presents a resilient approach to resourcing the Air Force technical training process based on best practices in the supply chain literature. Organizations must assess their uncertainty—which includes requirements, customer demand, and supplier uncertainty—and consider the appropriate resourcing method. Generally, resourcing approaches can be categorized as either proactive, where additional resources are secured beyond what is needed to meet current requirements, or reactive, which relies on swift processes to secure resources once requirement changes are observed. For organizations to have resilient resourcing, they must achieve a balance between the two. To achieve this balance, best practices recommend assessing uncertainty, determining resource capacity and options to increase capacity, and making resourcing decisions given capabilities, cost, and risk aversion.

We presented a five-step resilient resourcing process specific to the Air Force technical training process: (1) categorize pipelines, (2) quantify resourcing bounds, (3) estimate probability for requirement changes, (4) estimate effects of potential resourcing strategies, and (5) choose a resourcing strategy. Steps 1, 2, and 3 are information-gathering steps to identify which AFSCs need to respond to potential requirement changes, the magnitude and timeliness of adding additional resources, and the possibility of requirement changes. Step 4 is an analysis and modeling step where the information gathered is combined and hypothetical resourcing approaches are modeled to estimate effects on the number of graduates, students awaiting training, and other performance metrics. In Step 5, the resourcing approach decision is made considering the information obtained and modeled in the prior steps along with external information on cost and desirability to add additional resources.

Four related recommendations will directly support the resilient resourcing approach:

- *Develop a capacity visibility capability where capacity estimates are accurate, updated regularly, and include feasible timelines and magnitude of available additional resources.* Capacity should be readily accessible to all relevant organizations including Headquarters Air Force, AETC, the 2AF, training squadrons, and schoolhouses. It is important to have this transparency both during the initial planning process and when requirement changes occur. This capability will enable Step 2.
- *Communicate potential requirement uncertainties in addition to a single planning target.* These uncertainties can be expressed as ranges around the PGL quantities and can be used to proactively assess whether additional resources should be allocated and on what timeline to meet potential requirement changes. This concept is presented in Step 3.
- *Re-create technical training modeling capabilities.* Models can evaluate the pipeline effect of resourcing strategies given potential requirement changes. These models can be used to manage leadership expectations on response timeliness and can be used to inform the proactive versus reactive resourcing decision. This is Step 4.

- Consider creating a single office that is responsible for gathering all information, modeling the effects given resource feasibility and potential requirement changes, and making the decision on the appropriate resourcing approach. Having a single office will centralize information collection and enable informed decisions regarding a resourcing approach.[24]

---

[24] In the U.S. Navy training process, the Navy has the Production Management Office that provides all of these functions. The Air Force does not have an analogous office.

Part IV. An Approach for Developing an Agile Technical Training Instructor Mix

Robert Corsi, Darrell D. Jones, and Lisa M. Harrington

13. The Case for Building a Flexible Instructor Approach

According to Air Education and Training Command (AETC)'s recent Continuum of Learning Vision, Air Force training processes lack the "agility" to meet the operational demands of the 21st century (AETC, 2017). To rectify this situation, AETC's future vision encompasses a training environment that is agile and learning-centered, using on-demand and on-command courses, blended learning, and rapidly updated and relevant content. This new training vision requires not only a complete review of how content is developed and delivered but also a complete review of what category of instructors are best suited to deliver the training. The challenge is daunting: AETC must produce trained airmen at the required skill levels in a system that can rapidly react to changes in the number of students without compromising training quality and within the budget approved for initial skills training.

This section identifies and evaluates alternative instructor options, or combinations, that would allow AETC to quickly react to changing student requirements using a new approach to instructor resourcing. Of course, the number of instructors alone does not determine training capacity. Instructors, training devices, classroom availability, and support facilities are interrelated to ensure production (as noted in Part III's discussion on the interplay of binding resources). However, in discussions with the research team, stakeholders across the Air Force repeatedly pointed to significant challenges with providing enough qualified instructors at the right place and at the right time to support technical training requirements. These challenges will be even more pronounced in the agile, on-demand training environment that AETC envisions for the future with such features as noncollocated instructors, blended learning, proficiency advancement, and other flexible approaches to training.

End strength (the total number of Air Force members authorized by Congress) can be a constantly shifting target. Unexpected force structure increases or reductions, driven by mission changes, conflict, or the threat of conflict, all drive perturbations in force management and changes in the number of initial skills graduates that are needed in each career field. Although programmatic changes are part of the end strength and training calculus, the Air Force has yet to develop a system that is flexible enough in the short-term to accommodate changes in initial skills training requirements without significant lead time. Attempts to adjust training capacity, or discover what adjustments might be possible, have become excessively cumbersome staffing exercises.

An important part of developing a more agile system for initial skills training is to design a flexible instructor pool that enables AETC to respond to changes in the required number of graduates in specific skills, during the year of execution if necessary, without the traditional lengthy approach of adding requisite instructors (requesting approval for additional manpower, finding the right instructor, qualifying the instructor for duty, and transferring the new instructor to the training base). By adopting a more flexible approach to instructor manning, AETC could surge or restrict production depending on the needs of the Air Force.

The current process for calculating the required number of AETC technical training instructors is model driven and dependent on (among other lesser inputs) the planned number of enlisted and officer accessions. Based on our discussions with AETC staff, the assumptions in the current model may need to be revisited, since the required number of trained personnel has become more dynamic, with Headquarters Air Force passing down changes in numbers to AETC just before or even within the year of execution as discussed in other sections. In addition, new methods of training delivery will require new calculations, since some delivery methods require more or less instructor man-hours.

Under the approach currently in place, the Air Force spends a significant amount of time during each programming cycle allocating exactly the correct amount of resources to meet student training requirements, no more and no less. Although it is understood at the Corporate Structure level (a deliberative group made up of senior officers who assist in the programming and budgeting process for the Air Force) that there needs to be increases in the number of instructors when there are increases in accessions, in practice, this happens only in the aggregate. However, in the individual training pipelines, there are challenges to identifying the specific instructor positions and individuals required for each AFSC, grade, and skill. Because of the inventory dynamics and priorities throughout the Air Force, the reality is that AETC never has all the instructors that they require—and especially not in times of accession increases.

The traditional AETC instructor at initial skills training is an active-duty officer or enlisted member assigned to an initial skills training location for a standard tour. Their tour length can vary but most are three or four years. Enlisted members are nominated by their commanders under the DSD assignment program, a program designed to pre-identify high-quality service members for important institutional duty assignments such as instructors or recruiters. Once selected for the DSD pool and chosen as an instructor, the individual must attend the Academic Instructor Course before they are qualified to begin instructor duty. The entire process, from nomination for DSD to assignment at the training base, can take 18 months, and selectees for instructor duty are not available until well into the next fiscal year. Although not part of the DSD process, officer instructors are normally identified a year in advance by their functional career field teams and endorsed by their senior functional leadership.

Currently, these AETC technical training instructors are not shielded from deployments. During our discussions with technical training instructors and their leadership, we found, on one hand, that the selection of instructors for deployments receives mixed reviews. Individual instructors understand that deployments can help them progress in their careers and, depending on the AFSC, provide real-world experience they can bring back to the classroom. On the other hand, the training squadron must often piece together solutions for dealing with the loss of the deployed member—classes are doubled up, leadership and administrative personnel not normally assigned to training provide instruction, and, in some cases, active component and reserve component personnel are temporarily assigned to backfill the instructor. Leaders at technical training schoolhouses report that deploying instructors may negatively affect the quality of training. If the Air Force continues with this practice, then a more systematic way of providing substitute instructors could be helpful both to the production of graduates and the quality of training.

As an alternative, there is an opportunity to increase training flexibility by creating a pool of ready instructors for a specific technical training pipeline—an approach that can help the schoolhouses better react to changes in training requirements. With built in flexibility, AETC and the First Air Force could focus only on major programmatic changes to end strength and training requirements. This would allow major resource changes to be addressed during Program Objective Memorandum (POM) deliberations, giving AETC a compelling case to justify resource allocations from the Air Force Corporate Structure.[25] A flexible instructor pool—drawing on a mix of active duty, reserve component, civilians, and contractors—would enable AETC to react more quickly to changes in training requirements. Solid planning in advance of the POM cycle, by course, that includes an optimal mix of instructors, could expand capability and significantly reduce the challenges associated with requirement changes in the execution year.

To investigate the potential for a flexible pool of instructors, a review of the options is required. The next section describes some of the many options available to AETC in filling instructor positions at the various school houses, while the following one outlines a method for how these options might be assessed when building a pool of instructors. Most likely, the best fit will be a combination of the options presented to arrive at a flexible instructor pool that suits the proactive resourcing needs of an AFSC technical training pipeline.

---

[25] A *POM* is a recommendation from the services to the Office of the Secretary of Defense concerning how they plan to allocate resources to meet the Service Program Guidance and Defense Planning Guidance. The POM presents the services' proposal on how they will balance their allocation of available resources. The POM includes an analysis of missions, objectives, alternative methods to accomplish objectives, and allocation of resources.

14. Instructor Pool Elements

AETC could consider a range of potential sources of instructors to create a ready pool to support their training requirements. Policy, and perhaps even statutory, changes may be required to be able to access these potential instructors. This section describes possible alternatives to the current practice of employing permanently assigned active duty instructors. We gathered these potential options from discussions with those involved in technical training policymaking, resourcing, and management at all levels of the technical training pipeline as a part of the research performed in the preceding parts of this document and in prior work (Harrington et al., 2017). In the next section (15), we demonstrate one systematic method for assessing each of these options to characterize when each might best be used. In all cases, changes should be accompanied by systematic assessments to ensure that there is no loss of quality and, if so, there is enough flexibility to ensure poor performing individuals or groups can be removed.

A. AETC Assigned Active Duty Instructors

The majority of technical training instructors are currently active-duty members. As outlined previously, there can be significant lead times in gaining approval for additional instructors and then selecting, training, and relocating an individual.

B. Retired Active Duty Instructors (Limited Recall)

All services have the ability to recall retired members to active duty in unusual circumstances when their skill is critically needed; however, there is a restriction of no more than 1,000 serving at any one time (DoD-wide) (U.S. Code, 2017). Retired service members who are still qualified to be instructors could be recalled to active duty without extensive retraining to meet critical shortages in the instructor force. However, by design, recalling members to active duty is not easy and requires Secretary of the Air Force approval. Recalled members count against active-duty end strength and therefore take the place of an active-duty member. In addition, these individuals would have to be volunteers. If the recalled member was not in the local area of the training base, the Air Force would incur an additional cost for a military move. To make this option responsive, the Air Force would require a system to identify and maintain a database of the retired members from critically manned specialties and gauge their interest in such a program—especially individuals located near initial skills training bases.[26] Without such a

---

[26] This identification could be accomplished by a continual survey of retired members or a more targeted approach restricted to career fields with greater instructor needs and assessing whether any retiring airmen would be interested in teaching in the future.

tracking system, volunteers would have to be solicited whenever instructors were needed, and the lead time for publicity and evaluation of interested individuals could reduce the flexibility to rapidly surge. If, for some reason, there was a critical long-term shortage for instructors with a particular expertise, individuals could be recalled for longer periods of time (for up to three years) and could provide stability for a particular training pipeline.

C. Prior-Service Instructors (Left Active Duty, Not Retired)

Not everyone who leaves the military retires. Many individuals serve one or two enlistments or fulfill their initial active-duty service commitment and obtain employment in the civil sector. Qualified instructors who leave active duty to go to the private sector can volunteer to return to active duty. Depending on the length of time since they served, these individuals may or may not need refresher training in their specialty. Separated former AETC instructors would not require attendance at the academic instructor course, which reduces the time required to bring them back to active duty and get them into the classroom to conduct training. Individuals would have to be volunteers and would most likely have expertise in specialties that are critically manned or possess unique knowledge and skills in high demand within the service. Like retired recall, these returning individuals would count against end strength; however, unlike the retired recall, individuals who leave active duty and then return are not limited by the number of years they can serve. This program has been used with success in the past to address shortages in aviators and other critical skills.

D. Currently Employed Air Force Civilians

Military personnel are not the only people who perform instructor duty in AETC. Air Force civilians (civil servants) are often permanently assigned to instructor duty. They provide a stable core of instructors who have in-depth expertise in their specialty. Since they are not required to participate in other military activities that reduce their availability, Air Force civilian instructors are more available to provide instruction and are not subject to deployments. However, Air Force civilian instructors can remain in positions for extended periods of time and their experience can become dated in some career fields (e.g., cyber), although may do so less in others (e.g., logistics). Although military manpower is "free" when allocated to AETC, Air Force civilian manpower must be funded by operations and maintenance funding, which may not be available when required. AETC and each training location would need to ensure that civilian funding is available for these instructors. There can be significant time delays associated with hiring a new civilian employee, but reassigning a current civilian employee could reduce the time for an individual to begin instructing.

E. Civilians with Instructor Experience (Retired Annuitants)

Like the military, Air Force civilian employees serve the government for extended periods of time and retire from federal service. These former Air Force civilian employees, who were qualified instructors at the time of their retirement, could return to Air Force civilian status under the retired annuitant program. Traditionally, these civilians return to service for a short duration (normally one to two years) to fill a critical Air Force civilian vacancy. They could return, given funding availability, to help fill a critical instructor need.

F. Guard and Reserve (Active Guard and Reserve)

Like active-duty military, full-time guard or reserve members can serve as instructors and can be funded by either active component appropriations or reserve component appropriations, depending on their unit of assignment. The use of active guard and reserve (AGR) individuals in full-time instructor positions requires close cooperation with the guard and reserves because of the impact on funding levels in the reserve component. In addition, under current statutes, their primary responsibilities should be in support of part-time reserve component members. Normally, the number of reserve component instructors is based on the number of active reserve component students in the pipeline. These instructors provide an increased level of stability, as active reserve component members traditionally enjoy more stability and less permanent change of station than the active force.

G. Guard and Reserve (Traditional Reservists)

Traditional reservists typically serve for four drill periods over two days each month and for two weeks of training at other times of the year. Reserve component members can be called to serve for extended periods of active duty under several authorities. They can serve for periods of time to fill temporary critical requirements using active component–funded man-days. The majority of these reserve component members transitioned from active duty, and some filled instructor duties previously in their careers. Additionally, depending on their civilian expertise, these individuals could already be trained in highly technical fields in which the Air Force needs current expertise. These individuals could serve as instructors again, for an agreed-on length of time, with agreement from the reserve component leadership.

H. Guard and Reserve (Limited Active Duty Recall)

Traditional reservists can be recalled to active duty when a critical shortage exists. Once these individuals are recalled, they are counted against overall Air Force and major command manpower allocations as any other active-duty person. AETC would need to have vacant active-duty positions for those recalled. These members could require instructor training and a military move may be involved. Once recalled, these members would be treated as active duty in all aspects and could even be deployed during the period of recall. Normally, individuals are recalled for no more than three years.

I. Prior Active-Duty Instructors (Unit Assigned)

Active military members are on call 24/7 and can be assigned temporarily to fill any need. Instructor requirements are no different. Active-duty military personnel assigned to positions outside of the training environment with instructor experience could be assigned temporarily to fill critical needs at the schoolhouse; this would likely impact the unit where the individual is permanently assigned. This arrangement would have to be a short-term augmentation of AETC instructors and would require significant justification to prioritize one mission (training) over another (day-to-day operations). This option would require AETC to have such a compelling need that their mission would take precedence over the member's current unit.

J. Contractors

The Air Force often uses contractors to execute specific tasks when it is determined that this approach is more advantageous, whether in wartime or peacetime. Contractors allow the military access to capability without the responsibility for upkeep and sustainment. However, contractors are not without their own challenges. Unless this was planned in advance during the planning cycle and incorporated in AETC's POM, contracting monies could be very limited, and AETC would have to have the contract ceiling authority.

K. AETC Overmanning

In industry, companies may use production employees as a source for instructors. By providing instructor training to employees whose primary job is operational, companies can develop a ready pool of instructor-qualified employees to meet surge demands. AETC could adopt a similar program and assign additional manpower to technical training locations. These additional assets could be performing nontraining, operational duties at a given location when the number of permanently assigned instructors is adequate, but quickly pivot to classroom instruction during increased demand.

This option would be a significant departure from current practices and would require an increase in manpower allocated to the base-level and operational missions at technical training locations and a commitment to maintaining this additional instructor capacity even if it is not always in use. Units would enjoy the increased manning until their members are tapped to meet a training requirement. Individuals who participate in these positions would need to be properly vetted, and incentives may need to be offered. Units would need to manage the disruption caused by the individual departing for instructor duty.

These instructor-ready personnel could also be assigned to locations near technical training bases and be ready to respond on short notice when needed. Both civilian and military personnel could participate. For this option to be viable and to allow AETC to do advanced planning, some level of increased military and civilian end strengths would be required and funded in the POM process. AETC may decide that some full-time manpower can be given up to have these less expensive "on-call" instructors available as needed.

L. Adjunct Military and Civilian Instructors

The options just described for creating a flexible instructor pool involve assigning individuals to instructor duty—either permanently or temporarily away from their primary duty (for active component, Air Force civilian employees, and reserve component). However, adjunct or part-time instructors, drawing on college and university practices (discussed in Part II of this volume), are a viable solution to examine. Collocated or noncollocated using online synchronous and asynchronous distance learning, individual members could volunteer to serve as instructors in their nonduty hours. Individuals who recently served as AETC instructors could especially be targeted. This approach is analogous to a service member agreeing to teach as an adjunct instructor for an evening or weekend course for a college or university, except, in this case, they are instructing an AETC course. Because these courses are often broken into modules during the day into blocks of a couple of hours, individuals could instruct during their nonduty time (even easier for asynchronous courses). The impact on their current unit mission should be minimal, similar to the case for civilians.

Implementation considerations include the need to track volunteers and their specific skills and instructor qualifications. In addition, there would likely need to be methods for incentivizing and paying these instructors with bonuses or stipends. Anyone serving today could qualify under this option (active and reserve component on active duty and civilians employed by either the active or reserve component). Reserve component members not serving on active duty could participate in the option outlined below.

M. Nongovernment Adjunct Instructors

The options described earlier involve those currently employed by the government (either military or Air Force civilians) who could also serve as adjunct instructors collocated or noncollocated using online synchronous and asynchronous distance learning. However, this option focuses on people with expertise to become instructors or return to instructor duties but who are not employed by the government. This option is similar to the contracting option J, but these contracts would be with individuals (vice companies) and for on-call services to fill increases in the required number of graduates from a specific course or for short-lived surge requirements.

These individuals could come from a wide range of categories, such as former military, former civilians, or retired civilians or military. In addition, uniquely qualified subject-matter experts from industry could also participate. This option could entail either standing contracts for on-call services or short-term agreements established quickly.

15. Instructor Suitability Evaluation

Criteria must be established to evaluate the appropriateness of different instructor options for different situations that may occur in technical training pipelines. Various characteristics including cost, flexibility, quality, stability, surge capability, planning lead time, short-term technical ability, unit impact, and end strength impact all must be considered to determine how best to meet instructor requirements. Table 15.1 gives an explanation of each criterion, with a color-coded evaluation scheme: green (positive or neutral impact), yellow (moderate impact), or red (high impact). We do not mean to suggest that this list of criteria is complete or that the color-coded assessments listed are necessarily appropriate or fixed. Our goal here is to outline a systematic method for how decisions for implementing a pool of instructors should be approached—that is, with a transparent and methodical set of criteria against an agreed-on assessment scheme. Using this type of approach, the instructor pool can be constructed appropriately for the needs of a particular technical training pipeline.

Table 15.1. Option Assessment Criteria

Criterion	Color-Coded Assessment
Cost: Incremental costs, fully burdened, when compared with the current mix of active duty and civilian instructors	Green = neutral to current baseline Yellow = slight cost increase over baseline Red = high cost to implement
Flexibility: Ability to adjust instructor levels to meet changing demands	Green = high degree of flexibility to adjust instructor mix Yellow = some flexibility Red = minimal to no flexibility outside of the normal programming process
Quality: Degree to which instructor quality could be maintained/assured	Green = more direct control over instructor quality at schoolhouse Yellow = less control over quality Red = little control of instructor quality and use of nontraditional instructors
Stability: Control over the instructor asset	Green = AETC owns the instructor asset, and the instructor is formally dedicated to instructor duty Yellow = AETC has some control but may be dependent on other organizations to use the asset Red = little control in maintaining the instructor asset for long periods of time
Surge capable: Ability to adjust in real time to meet emerging requirement	Green = instructor option can be adjusted in real time (3–6 months) to deliver an emerging requirement Yellow = instructor option can be adjusted in real time (6 months to 1 year) to deliver an emerging requirement Red = little surge capability; dependent on the POM process

Planning lead time: Time to ensure that option could be executed when needed	Green = could be employed during execution year with little to no prior planning Yellow = some prior planning necessary and possible need to include in budget submission Red = totally dependent on competing in the POM process
Short-term technical: Ability to provide almost immediate on-call capability	Green = capability exists immediately Yellow = capability exists but is not immediately available and could take up to 6 months Red = longer lead time regardless of the source
Unit impact: Impact to units not assigned to AETC or assigned to AETC in non-tech training missions	Green = minimum impact to units not assigned to AETC Yellow = some impact to units for a short period of time Red = significant unit impact
End strength impact: Growing end strengths may be required	Green = no military or civilian end strength impact Yellow = indicates that there could be a mix change (among active, guard, reserve, or civilian), but no change to end strength totals Red = requires an increase to required end strength

Table 15.2 shows the application of the criteria to each of the instructor options. Following the table is brief discussion of the assessment for each of the options giving some rationale. As stated previously, these options and assessments are not definitive; what we present here is a notional assessment providing an example of the process AETC might undertake to establish a flexible instructor pool. As any such assessment might vary by schoolhouse or even course, we would recommend teams of experienced course planners, contracting officers, and quality-assurance managers provide input into these assessments, as many of them operate at this level and have ready access to the relevant data.

Table 15.2. Assessment of Instructor Option Suitability by Criteria

Criteria	Instructor Option												
	A	B	C	D	E	F	G	H	I	J	K	L	M
Cost	R	Y	R	G	G	R	R	Y	G	G	G	Y	G
Flexibility	R	G	G	Y	Y	Y	G	G	R	G	G	G	G
Instructor quality	G	G	G	G	Y	Y	G	G	G	G	G	G	G
Stability	G	G	G	G	G	G	G	Y	G	G	G	R	R
Surge capable	R	R	R	R	G	G	G	G	G	G	G	G	G
Planning lead time	R	R	G	G	G	G	Y	Y	R	R	R	Y	G
Short-term technical	G	G	R	G	G	G	G	G	G	G	G	G	G
Unit impact	G	G	G	G	G	G	Y	Y	G	G	Y	G	G
End-strength impact	R	R	G	G	G	Y	Y	Y	G	G	G	G	G

NOTES: Letters in the second row correspond to the instructor option. Color of cells indicate how the criteria applies to an instructor option. Green = positive or neutral impact based on current method; yellow = some impact, potentially added cost; and red = high impact, high cost, long lead time.

A. *AETC assigned active-duty instructors*: This option is not conducive to meeting the short-term demand for an increase in instructors for initial skills training. Based on the forecasted number of students, each year the Air Force allocates instructors to AETC. Although this is not the highest-cost option, it is the option with the least flexibility. It is a very stable option and comes with rigorous instructor standards but is considered red on surge, requiring significant planning lead time and is dependent on the POM cycle. This is currently AETC's prime instructor model, and active-duty instructors will likely remain at the core of the instructor pool.
B. *Retired active-duty instructors (limited recall)*: Overall, this is a highly desirable option with great flexibility. There is some incremental cost, and the funded position must already exist. But instructor qualifications may need to be improved with extra training.
C. *Prior-service instructors (left active duty, not retired)*: Overall, while there is an end strength and cost impact when using this option, all other categories are rated high due to being able to leverage the prior instructor training of these individuals.
D. *Currently employed Air Force civilians (active)*: From a cost standpoint, this option represents the least costly of instructor options when needing to provide training for an increased number of accessions. A downside is that civilian employment levels tend to be a target when trying to reduce spending during budget deliberations.
E. *Civilians with instructor experience (retired annuitants)*: Overall, this is a strong option that rates high except on cost and flexibility. The flexibility limitation is due to civilian employment rules. Given employment rules, the Air Force hires civilians and grades their positions based on defined tasks included in their job description. Since retired annuitants have very defined skill sets that are needed in the short term and that the justification for bringing them on board is based on that skill set, there are significant limitations to use them outside of their job description.
F. *Guard and reserve (Active Guard and Reserve)*: Similar to the active duty in many respects, AGRs are also the most costly when all of the costs are included. Overall, this is one of the highest cost options, requiring planning lead time, and is not a good short-term solution with end strength implications. From a fully burdened cost standpoint, AGRs are the most costly when compared with active duty in the pricing models used in the planning and programming process. In addition, AGR end strengths are tightly managed by Congress and must be planned for and budgeted well in advance, which limits their ability to be used in the short term.
G. *Guard and reserve (traditional reservists)*: With the cost of TDY salary and other allowances, these members can be very costly in the short term. Planning for man-days must be considered during the planning cycle to be incorporated in POM deliberations and in the prioritization of man-day assets. This option is good for short-term, highly technical requirements. Overall, it is a very good surge option.
H. *Guard and reserve (limited active-duty recall)*: Overall, it is a very good surge option but does impact active component end strength, since the member is filling an active component position. It also depends on whether the reserve component must backfill the position their member leaves vacant.
I. *Prior active-duty instructors (unit assigned)*: Although this is one of the least-expensive options for short-term requirements, it would be hard to identify the effect on the current unit's impact. This option is good for short-term, execution-year critical mission needs. Overall, some limitations exist for flexibility, stability, and unit impact.

J. *Contractors*: Expect contractors to be the most expensive nonorganic option based on historical contractor full-time equivalent cost when procuring professional services, but it may be the only option to meet a short-term technical training need. Overall, this option is costly and requires planning lead time, but it is positive in most other aspects (quality control can be challenging if contracts are not written to clearly remove poor performers). Since there is a current prohibition against outsourcing under Office of Management and Budget Circular A-76 in the National Defense Authorization Act, which has been continued since 2008 through FY 2017 for the current organic workload, this would pertain to meeting a new workload requirement, and AETC would need to have the contract ceiling. However, there is ongoing discussion in the Congress to remove the prohibition on A-76 competitions for 2018.
K. *AETC overmanning*: Overall, overmanning is costly with end strength implications—unit cohesion could be affected, and instructor quality could be highly varied. Because of the end strength effects, lead time is needed, and it is not a good option for short-term technical demands. However, this option does offer flexibility and surge capability.
L. *Adjunct military and civilian instructors:* Using adjunct instructors would be a relatively low-cost option. The Air Force currently uses special pays and incentives to encourage members to volunteer for various types of duty. This option requires advertisement of which skills are needed and a database to keep an accurate track of those willing serve as adjunct instructors. This option rates high for flexibility and low for impact on unit mission and end strength.
M. *Nongovernment adjunct instructors*: This option is the most flexible and lowest cost of all the options and offers the greatest payback. One concern is predicting the number of individuals available with appropriate expertise and qualifications. This option takes advantage of the experience of former or retired military without having to return them to active duty.

Creating an Instructor Pool Using the Assessment Criteria

Not all of the options for populating an instructor pool are currently available. There are challenges remaining for the implementation of several of the options. The goal is to identify a set of options for creating a pool of instructors that matches the needs of a particular training pipeline. Some pipelines have so little variability in the number of students that flow through annually that creating a pool of available instructors would not be an efficient resourcing solution. On the other hand, the student flow for some AFSCs is highly variable. Some AFSCs are targeted often for an increased number of students when end strength is increased, and the number of technical training graduates must change with relatively little advanced planning.

The options and assessment criteria described here are the first steps toward developing a strategy for creating a pool of available instructors for a particular AFSC. Individual characteristics of a training pipeline (the length and number of courses, washback rate, use of classified systems), the location of the schoolhouse, the manpower environment, the operational implications of shortages, and many other factors will need to be considered.

A Notional Example

To understand how AETC might create a pool of instructors to augment permanent instructors when needed, imagine a notional scenario where it is highly likely that the requirements for intelligence officers will increase because of tensions on the Korean peninsula. The Air Force decides to proactively resource the capacity to train Intelligence Officer (AFSC 14NX) production. For this example, they predict a high probability of a 10 percent increase in the number of required graduates in the next two fiscal years. Because the traditional approach to acquire more full-time active-duty instructors would not be responsive enough (requesting approval for additional manpower, finding the right instructor, qualifying the instructor for duty, and transferring the new instructor to the training base), the Air Force and AETC leadership decides to create a pool of on-call instructors ready to provide instruction to support the increase in graduates. The Air Force and AETC determine that they want the pool of officer intelligence instructors to have the following prioritized characteristics:

- It is already six months into the fiscal year, so AETC has only six months to build the pool.
- Instructors should be of the highest quality since intelligence officer graduates should be "mission ready" as soon as possible after reaching their permanent duty stations.
- There should be a high degree of flexibility to adjust the number of instructors being used from the pool.
- Given the importance of the intelligence mission at the time, the Air Force and AETC want to minimize the impact to intelligence units across the active and reserve components.

Reviewing the assessment in Table 15.2, several options satisfy the surge requirement with only six months to implement. Four options are best suited to the characteristics above:

- contractors (option J)
- separated prior instructors and civilians with instructor experience (options C and E)
- nongovernmental adjunct instructors (option M).

Since cost is not rated as a high priority, AETC could contract with a company to provide a pool of instructors within the required timelines if the increase in the number of intelligence officer graduates actually materializes. However, the fiscally prudent approach would be to investigate the number of individuals who are qualified and ready to serve as instructors from options C, E, and M. Options C, E, and M are some of the same individuals—the question is whether they are available to instruct full time (options C and E) or whether they are only available part time (option M). Contacting and tracking these individuals may pose a challenge.

The Air Force would put a call through all available multimedia channels—websites, advertisements in military-related publications military and veteran service organizations, social media, and others. Another call would go out to retired military and civilians through the official Air Force website (www.af.mil), nonofficial Air Force–related websites or blog postings, and through professional organizations (e.g., the Air Force Associations and the U.S. Air Force

Association of Graduates) that provide an online location for people to express interest and availability and sign up so the Air Force can do initial screening. Based on this initial screening, individuals are selected to be included in the pool to be tapped when and if additional courses are added for intelligence officer training. Depending on the number of qualified individuals and their availability, AETC could then decide if additional contractor instructors are required for the pool.

Recommendations

In this assessment, we examined instructor options in isolation against a set of assessment criteria. But any instructor option must be considered within the broader construct of how AETC will implement their blended learning concept and how enabling technology could significantly change the way course content is delivered and, in turn, the demand for instructors. We recommend that AETC use an approach such as the one demonstrated here to systematically investigate options for creating a pool of instructors when it is determined that a proactive resourcing strategy is appropriate (see Part III for further discussion of resourcing strategies) or when immediate needs for instructors increase. Such an approach should

- Identify potential sources of instructors.
- Ensure assessment criteria represent the characteristics desired of such a pool.
- Systematically identify barriers to using particular categories of manpower for an instructor pool and develop policies to mitigate.
- Develop pilot programs for promising sources of instructor manpower for a small set of AFSCs with critical shortages. Ensure results are fed back into the process for identifying and assessing options.

Part V. Concluding Thoughts

Bart E. Bennett, Kathleen Reedy, and Lisa M. Harrington

16. The Benefits of Flexibility for Improving Technical Training Efficiency

What benefits could AETC obtain by introducing more robustness and agility into the planning and execution of courses? We explored this question by examining some flexible resourcing options and their effect on one example benefit: the number of student-awaiting-training (SAT) days. SAT days represent a potential inefficiency in the technical training system, as airmen must wait before receiving essential training. This can have both a negative impact on readiness, since these airmen are delayed in reporting to their first assignments, and on trainee morale. As the basis for this analysis, we used the first course in the F-16 maintenance (AFSC 2A333M) series (course J3AQR2A333M 026B/C) called Maintenance Apprentice (F-16) Part 1. We used three years of data (2014, 2015, and 2016) and derived the annual number of SATs for each option to demonstrate the extent potential benefits might accumulate with alternative flexible management strategies.

The annual summary data is shown in Table 16.1. This includes:

- SATs on Jan 1: This is the number of students at the beginning of each year waiting to enroll in a class. These students could have either been in excess of the number of seats available in the last class of the prior year, arrived as new students after the last class of the prior year started, or had washed back after the last class of the prior year started.
- New Student Arrivals: These are students that are taking the course for the very first time. They include active-duty airmen from BMT, Air National Guardsmen or Reservists, or those cross flowing from another career field.
- Washbacks: As explained earlier, these are students that fail a curriculum block during the training sequence but are permitted to repeat the course beginning with that block when a seat becomes available in a subsequent class.
- Total Seats Required by Students: This is the sum of SATs on Jan 1, New Student Arrivals, and Washbacks.
- Planned Seats: This is the annual number of class seats supplied during the year.

Our first observation is that the SATs on January 1 are not just a simple difference between the annual total seats required by students and the planned seats the prior year. For example, since the required seats for 2014 were 447 and the planned seats were 455, it would appear that the SATs on January 1, 2015, should have been zero. However, four new students arrived from BMT right at the end of the year after the last scheduled class start date for the year. These students had to wait until the first class of the new year started, meaning that they were in SAT status on January 1, 2015.

This kind of mistiming between students ready to start a class and the start of a class is one cause for SAT days. It not only occurs at the end of the year (see the example illustrated earlier) but also throughout the year. Furthermore, in addition to students who arrive in advance of a course, students in excess of the class size, either because of delayed or unexpected entries or a larger number of washbacks than anticipated, will cause additional SATs to accumulate. In either case, SAT days will continue to accumulate until the occurrence of a class or classes with enough vacancies to clear the student backlog. Student arrival postponements when insufficient students are waiting can exacerbate the number of SATs by leaving a class seat vacant while requiring an unplanned seat at a later date. The overall result can be a backlog of students waiting training and a large number of SAT days that accumulate. The purpose of this section is to demonstrate how SAT days might be reduced if more flexible course management were instituted.[27]

To estimate the number of SAT days, we compared the student arrival schedule with the planned class schedule and class sizes. Although the data system is capable of recording SAT days, we found the data are not entered with sufficient discipline to reliably compute total SAT days.[28] These data would not allow us to reliably determine the cumulative annual number of SAT days—the primary measure we wished to use for this comparison.

It was therefore necessary for this demonstration to simulate the assignment of students to classes in order to estimate SAT days. We use the simple assumptions that students are placed in class on a strict first-come, first-served basis, with all students arriving when expected and none delayed or canceled. These assumptions tend to fill classes as efficiently and as full as possible, leaving as few seats as possible empty and creating fewer SATs than are realized under less-optimistic conditions.[29] Estimated outcomes under these assumptions, denoted as the "idealized" case, are shown in Table 16.2. Four measures are shown in this table to summarize the estimated SATs occurring during each year.

Although the total number of planned seats in 2014 and 2015 exceeded the total number of students requiring training (from Table 16.1), the cumulative SAT days were significantly large. In 2016, when planned seats were less than those required, the cumulative SAT days ballooned even larger. A major reason for this is because of the mistiming of student arrivals and class starts. Table 16.2 shows that the average daily SATs in both 2014 and 2015 are each equivalent to a full class for this course. On average, in 2016, the equivalent of two full classes of students waited every day all year long. But this is just the average. Some days, the number waiting was sometimes less and sometimes significantly more. Looking first at a measure of when those waiting was less, only 11 percent of the time in 2014 was there ever no one waiting. In other

---

[27] This is not a thorough analysis but merely a demonstration of the potential benefits. A thorough analysis would consider implementation details including a cost-benefit analysis comparing the benefits with the costs for implementing more flexibility. The analysis would need to be done on a course-by-course basis to include the unique characteristics of each course.

[28] Students can be taken out of training status because of illness, family, disciplinary, or other reasons.

Although we do not know the methods used by course managers to fill seats, we found that seating as many students as were available, up to the planned class size, closely approximated the observed annual number of seated students. Less-efficient class filling will occur for a number of reasons, particularly seats left unfilled by delays in planned student arrivals, by illness, or by other circumstances.

words, 89 percent of the time, a backlog of at least one student occurred. The similar result for 2015 and 2016 were even worse with the backlog never clearing in 2016. As for the maximum number of students that waited during the year, 2014 and 2015 had as many as three full class-load equivalences waiting, while 2016 had as many as five waiting.

Table 16.1. Data for Maintenance Apprentice (F-16) Part 1

Year	SATs on Jan 1	New Student Arrivals	Washbacks	Total Seats Required by Students	Planned Seats
2014	9	366	72	447	455
2015	4	405	58	467	472
2016	9	505	81	595	581

SOURCE: Transactional data of Enlisted Skills Training students from the AETC Decision Support System (ADSS) database obtained May 2017.

Table 16.2. Estimated Outcomes for the Idealized Case for the Maintenance Apprentice (F-16) Part 1 Course

Year	Cumulative SAT Days	Daily Average SATs	Percentage of Day with Zero SATs	Maximum SATs
2014	2,896	7.9	11.0	22
2015	3,230	8.8	4.4	24
2016	6,390	17.5	0.0	42

SOURCE: Author's calculations on transactional data of Enlisted Skills Training students from the ADSS, obtained May 2017.

What if alternative, more flexible class-management options had been used, particularly in 2016? Could the number of SAT days have been reduced? Rather than a comprehensive analysis, we desired to explore a small number of relatively simple management options to demonstrate the plausibility of significantly reducing SAT days by management intervention. We chose three options that changed either the allowable class size, the planned schedule of classes, or both. We have given these approaches a simple short name and explained how flexibility was introduced in order to reduce the mistiming of student arrivals and available class seats as follows:

- *Flex size*: This approach retains the planned class schedule but allows some flexibility in the number of seats available in each class. Assuming a normal maximum class size of eight students, we allow classes to swing from a low of "half" classes (of four students) up to "one and a half" classes (of 12 students). Although some class sizes in the actual data surged beyond this size (a couple of classes had 13 or 14 students), we limit the class size to 12. Within these class sizes, we assume that only a single set of class resources (e.g., instructors, training devices, class rooms) would be needed. This option packs classes fuller than desired but could be implemented to reduce a backlog.[30]

- *Flex instructors*: This approach retains the planned class schedule but adds instructors (and therefore additional classes) when SATs reach a certain threshold. When four or more students wait to enter a class, we allow all the students waiting to be taken out of the SAT pool and seated in a new class, which will start *at the same time* as the class in the planned class schedule. Such an option would require a great deal of flexibility in the instructor pool and other resources to accommodate multiple simultaneous classes. It is likely that this option for seating students is unrealistic under current practice, however, it provides an approximation about the number of SAT days that are caused by the instructors serving as the primary binding resource.

- *Flex schedule*: In this approach, we potentially use the classes planned in the schedule but also allow additional classes to be held whenever there are at least four students in SAT status. These classes can start *at any time*, they do not need to start at the same time as the class(es) in the planned class schedule. We constrained the maximum class size to the "one and a half" limit used in the flex size option. So, for example, consider a day with no scheduled class and three students in SAT status from the prior day. If any students arrive, the minimum class threshold of four will be exceeded, and a new, unplanned class is started. If no students arrive, the three in SAT status carry over to the next day. Furthermore, if an insufficient number of students are available (fewer than four) on a day in which a class was planned in the schedule, the planned class will be canceled, and any students waiting are carried over to the next day.

All four cases (idealized and the three options) result in the selection of a different number of classes held and the sizes of the classes. Table 16.3 shows the number of class starts planned in the original class schedule and the number selected in the cases we considered. Under planned classes, the first value is the number of classes actually held followed by the number that were initially scheduled but later canceled.[31] For the purpose of our simulated cases, we permitted all of these planned class starts (all 71 in 2014 and not just the 64 that were actually held) to be used as candidate start dates in all four cases. The idealized case uses all of these candidate start dates in 2016 but requires one fewer than the actual held classes in 2014 and 2015. The extra class actually held is because of inefficiencies in seating students described above. Course offerings for flex size and flex instructors are generally reduced as the class sizes are allowed to be larger, particularly for few instructors, than in the idealized case. The number of classes for flex schedule is more mixed, with classes provided more closely with the arrival of students. As the total number of seats required by students increases from 447 to 467 to 595, the number of classes using flex scheduling increases from 55 to 60 to 71. The class starts in Table 16.3 include a range of class sizes from half classes (four) to classes that would need to be accommodated

---

[30] We realize that there is danger in suggesting that such a temporary surge in classroom capacity is possible. It would be far too easy to change the temporary surge capacity into the required capacity without fully considering the negative impact on student learning, instructor morale, and impact on instructional resources.

[31] It appears that these canceled class starts were not held due to insufficient numbers of students available to start the class.

with multiple simultaneous class offerings. The distribution of these sizes is shown for each case in Figure 16.1.

Table 16.3. Number of Class Starts for Seating Options
for the Maintenance Apprentice (F-16) Part 1 Course

Year	Planned Classes (held/cancelled)	Number of Class Starts			
		Idealized	Flex Size	Flex Instructor	Flex Schedule
2014	64/7	63	52	47	55
2015	57/6	56	52	44	60
2016	64/0	64	57	50	71

SOURCE: ADSS class schedule data and RAND-generated simulation results.

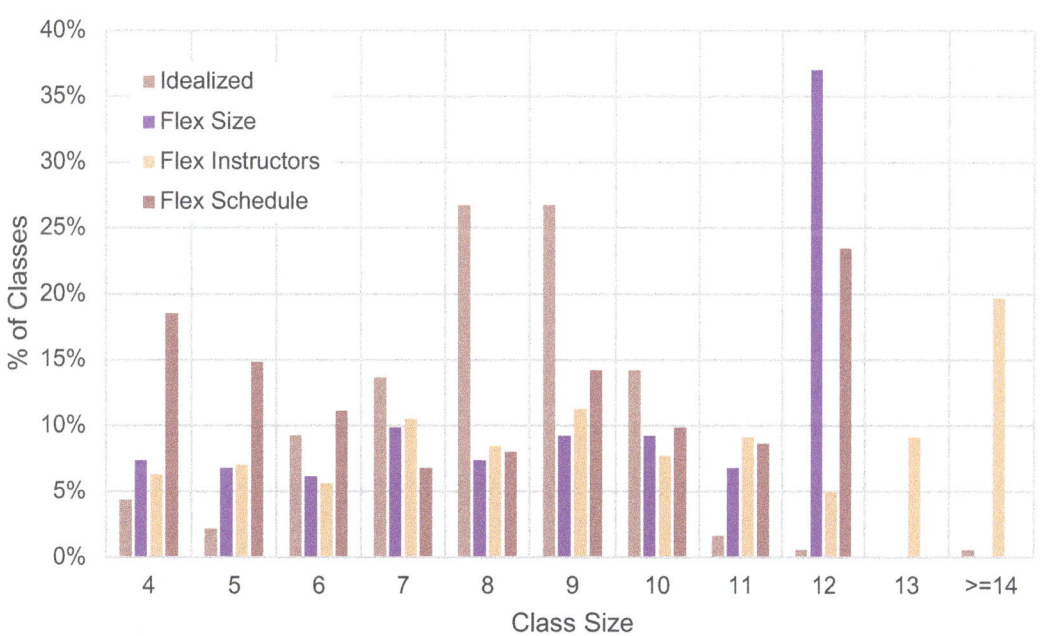

Figure 16.1. Distribution of Class Sizes for Seating Options
for the Maintenance Apprentice (F-16) Part 1 Course

SOURCE: ADSS class schedule data and RAND-generated simulation results.

Figure 16.2. Cumulative SAT Days for Seating Options for the Maintenance Apprentice (F-16) Part 1 Course

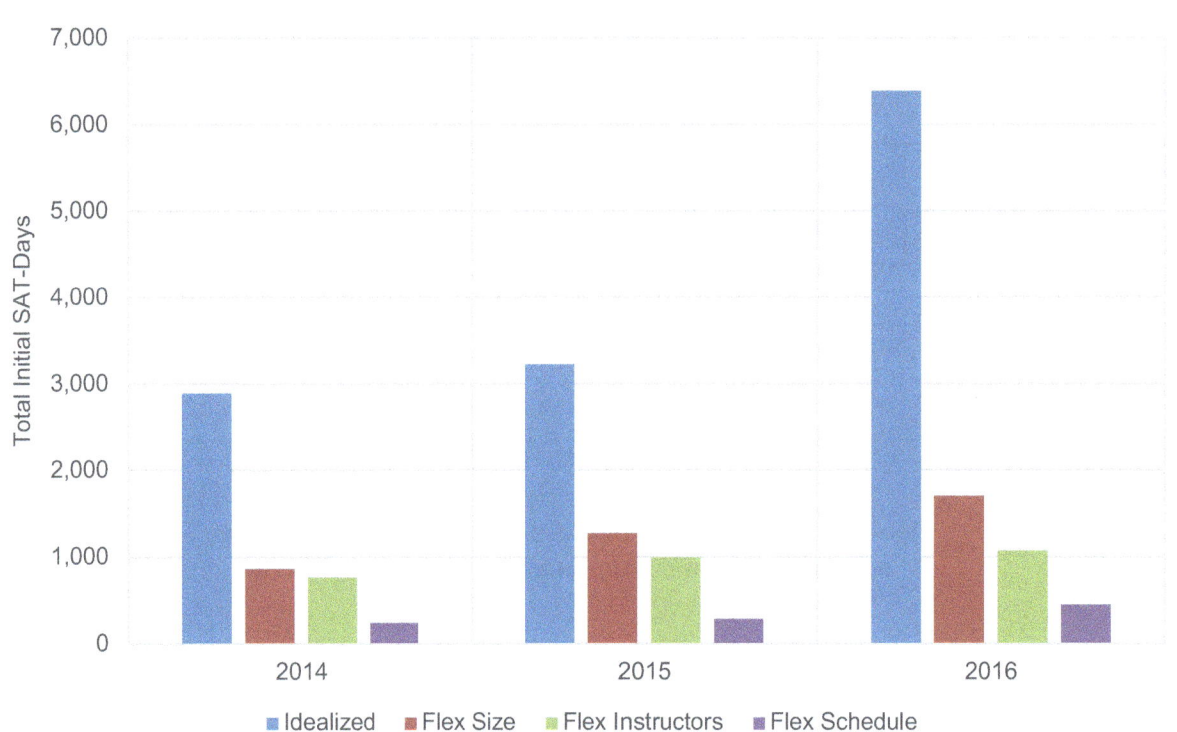

SOURCE: ADSS class schedule data and RAND-generated simulation results.

The most important result to consider, as shown in Figure 16.3, is the significant reduction in the cumulative number of SAT days each year that can be realized by employing these flexible resourcing alternatives. As the figure shows, flexing can have a dramatic impact on SAT days compared with the idealized case, which produces a very large number of SAT days, particularly in 2016, when more students needed to be seated in a class than the planned capacity.

- Allowing the *class size* to expand to 12 when sufficient students are waiting results in fewer classes taught but a dramatic increase in the number of classes with a class size of 12 students. This kind of surging, seen very occasionally in the actual data, might not be sustainable without a significant increase in resources.
- In the *flex instructors* case, the number of class starts is smallest, but the total number of students desiring to start at the same time is relatively large and would require conducting two to three simultaneous courses. The additional benefit from this is significantly less than the impact of just flexing the class size between four and 12 students. Although surging the class size without additional resources—such as instructors and training devices—is unrealistic, we can use this example to illustrate the impact on student wait days if the additional resources would be made available.
- Allowing *alterations to the schedule* and seating a class with up to 12 students when the number waiting exceeds four causes the greatest reduction in the number of SAT days. Surprisingly, the number of class offerings does not exceed the idealized case in 2014

and only marginally exceeds the idealized case in 2015 and 2016. These results are achieved by increasing the class size to 12 but not as frequently as the flex-size case. Interestingly, it is also achieved with a significant number of smaller class sizes.

These cases demonstrate the extremely large decreases in SAT days that could be achieved by more flexible class sizing and scheduling. Such changes would require dynamic access to additional resources, particularly instructors, in order to achieve at least some reduction in SAT days. Although the extremes shown in this example would be difficult to achieve, it does illustrate the potential benefits that could be achieved by greater robustness and agility in planning and resourcing.

17. Where Should the Air Force Go from Here?

Previous RAND work (Harrington et al., 2017) identified numerous opportunities for improved efficiencies in the Air Force technical training pipeline, particularly those that would offer the Air Force greater flexibility in responding to changing demands. Parts II–IV of this report have each explored different models and concepts that AETC can adopt, in whole or in part, to change how they currently operate in planning and executing technical training in order to increase efficiency. Each discussion is primarily focused on how to improve the adaptability of the system, particularly in response to rapidly changing requirements for how many and what type of airmen are being trained. From the comparative look at how colleges and universities manage similar processes, to the exploration of a supply chain management approach to planning, to the practical assessment of how to weigh options in an instructor mix, two major themes have emerged.

First, there is no one-size-fits-all model that will work across the whole of AETC. As noted in Part II, differences that may exist between AFSCs—such as course length and requirements to conduct lessons in person versus online—can impact decisions about how to employ adjunct-style instructors. Part III emphasizes that decisions about whether to adopt proactive or reactive resourcing strategies, particularly with respect to instructors, should be based on an assessment of the needs of each AFSC including expected demand variation. Part IV highlights the wide variety of options that AETC and schoolhouses might use to create a flexible instructor pool and emphasizes the need to assess each case separately when additional resources are needed on short notice. All of these approaches will be most effective if AETC makes most of its resourcing decisions at the AFSC and training squadron level. Doing so will optimize the use of resources and improve the pipeline's efficiency. Such decentralization must be accompanied by an accountability system that tracks costs and capacity and provides visibility into the schoolhouses to all management levels—wings, AETC Headquarters, and the Air Staff. At each management level, information needs to be accessible to permit better-informed decisionmaking and wise use of scarce resources.

Second, and perhaps more important, flexibility across the training pipeline is the key to improving planning and resourcing efficiency. The Air Force frequently advocates the importance of agility but, as we discuss in Part III, this term can have a broader and quite different range of meanings than the context in which it is often used by the Air Force. Supply-chain management principles suggest that the optimal resourcing strategy would be one that appropriately balances agility (as a reactive approach) with robustness (as a proactive approach). But it is important to know when and how to be able to appropriately find and alter that balance, which requires flexibility within the processes guiding resourcing and planning. Using adjunct-type instructors or creating a flexible instructor pool, as described in Parts II and IV, requires a similar need for flexibility. Without the flexibility to run different parts of the technical training pipeline in different ways or to conduct regular analysis that leads to adjustments to the process over time, current inefficiencies will continue to cost the Air Force time and money, as has been underscored in prior RAND research (Harrington et al., 2017).

Employing the Concepts Reviewed

Based on the concepts reviewed in this report, we offer a number of suggestions for how AETC can begin to implement these practices, as well as considerations for the use of metrics and measures to ensure Air Force technical training continues to produce quality graduates. These suggestions cover four main areas: how to gain efficiencies in planning using supply chain models, how to gain efficiencies in instructor mix, how to gain efficiencies in content delivery, and how to ensure quality is not being lost (or is actively being gained) through the implementation of new initiatives.

Planning

- **Categorize AFSCs by how reactive or proactive they should be.** Using the methods presented in Part III, determine which AFSCs would benefit most from a more reactive or proactive approach in determining instructor numbers.
- **Prioritize AFSCs within these categories.** Using Air Force priorities for manning, prioritize which AFSCs would make good candidates for proactive approaches. Because these approaches cost more (i.e., require having additional instructors on hand), a prioritized list will help determine how to best allocate resources.
- **Conduct a pilot to test the efficiencies gained by using reactive and proactive planning approaches.** Using the list developed, select a number of AFSCs designated by each approach and use that approach during the planning process. Conduct quantitative and qualitative assessments over a two- to three-year period to asses any efficiencies gained and whether to apply the model to all AFSCs.

Instructor Mix

- **Assess benefits and costs of instructor mixes currently employed in schoolhouses.** Identify AFSCs that currently use different instructor mixes and quantitatively and qualitatively examine the strengths and weaknesses for each mix, ascertain why certain mixes are more advantageous in certain AFSCs, and examine mixes' applicability to other AFSCs.
- **Conduct an in-depth assessment of the costs and benefits of different instructor mixes.** Using the notional rubric presented in Part IV, conduct a study to assess the strengths and weaknesses of the full contingent of instructor mix options to provide a basis of knowledge for future consideration.
- **Pilot the use of various mixes on select AFSCs.** Proactively select several AFSCs that are strong candidates for using different instructor mixes (based on a variety of availability of non–active duty instructors, sensitivity of course content, need for up-to-date operational experience, classroom versus hands-on content). Initially, use courses

that have low volatility in graduate requirements to ensure that the pilot will not disrupt the number and quality of graduates. Assess different mixes based on timeliness, cost, and quality of instruction, then apply to other AFSCs progressively.
- **For courses suited to online or blended learning, pilot the use of remote adjunct-like instructors.** In conjunction with an assessment of which technical training courses may be best suited for online or blended learning (see below), pilot and assess the quality of learning using instructors at remote locations using the options provided in the instructor mix assessment in Part IV.

Content Delivery

- **Develop and pilot alternative course delivery methods, such as online or blended learning:** Using input such as the need for classified or hands-on environments, identify courses or portions of courses suitable for alternative delivery methods and conduct a multi-year pilot to test for changes to the quantity or quality of education. If successful, identify further courses suitable for such methods.
- **Pilot competency-based learning in suitable courses.** Between the time of initial drafting and final publication, AETC has already implemented such a pilot in several courses. From there, AETC should carefully assess the qualitative and quantitative effects of such a model over several years and, if successful, should identify further courses to which to apply this model.

Quality Control

- Establish current and, if possible, historic baselines for technical training courses; compare the results of the recommended pilots and initiatives, or other initiatives identified by the Air Force, to that baseline to determine whether there has been a gain or loss in quality. Continue to monitor these metrics on a routine basis over time. Metrics should be both quantitative and qualitative. Although not an exhaustive list, some options to consider are
  - quantitative
    - graduate numbers in a fiscal year
    - student grades
    - average, median, and extreme high and low times to complete a course
    - annual numbers of students not in training and time spent not in training
    - rates of washbacks or washouts
  - qualitative
    - overall quality of the graduates (some AFSCs make attempts to assess this by conducting interviews with graduates' first commanders, but such assessments should be standardized and routinized)
    - student satisfaction with the courses
    - instructor satisfaction with the courses.

Conclusion

Flexibility in resourcing has a lot to offer in terms of improving AETC's efficiency in the nonrated technical training process, particularly in a continued environment of sometimes rapid changes to requirements for graduates. As the previous section demonstrated, the ability to readily flex class size, instructors, and schedules has the potential to significantly increase the production of graduates in a timely manner and, in doing so, reduce costs to the Air Force of students not in training. More importantly, a more efficient flow of students through the training pipeline reduces or eliminates unfilled positions in operational units and the potential decrement to readiness that manning shortfalls can cause.

Although making changes to class size and schedules can be readily handled at the training squadrons, changing the way that AETC resources and produces instructors is a more challenging endeavor. Thus, Parts II–IV of this report focus on justifications and alternative models that offer the Air Force suggestions for how such concepts might be employed. The flexibility to apply different resource strategies to each training pipeline, adopting lessons from colleges and universities to leverage technology and employ both permanent and part-time faculty, and systematically and routinely assessing the best mix of instructors are options AETC can use to improve its ability to rapidly adapt to changing requirements in a very uncertain world.

Appendix A. Statistics of Air Force Special Code Program Guidance Letter Changes

This appendix displays data tables on and categorizations of the Air Force Specialty Codes (AFSCs). The tables display:

- *AFSC*: Air Force Specialty Code
- *AFSC Description*: description of the AFSCs
- *Initial PGL (PGL Change) by Year*: For each year, the table elements show the initial PGL and, in the parentheses, the PGL change or difference with the final PGL. For example, if the initial PGL is 100, and it was changed by reducing 20, the table would display "100(–20)"
- *Avg Initial PGL*: the average PGL for the AFSC overall years for which we have data
- *Avg % Change*: the average over the years of the changes expressed as percentages of the initial PGL
- *% of Years >25% Change*: the percentage of all observed years that had an increase or a decrease change of more than 25 percent
- *Category*: a category to group AFSCs that have similar "Avg % Change" and "% of Years >25% Change."

The categories are defined as follows:
1. Average PGL change is within ±20 percent and less than 20 percent of the years had a greater than 25 percent increase or decrease.
2. Average PGL change is within ±20 percent and more than 20 percent of the years had a greater than 25 percent increase or decrease.
3. Average PGL change is greater than a 20 percent increase.
4. Average PGL change is greater than a 20 percent decrease.

Table A.1 shows Enlisted AFSC that are in Category 1. Table A.2 shows Enlisted AFSCs that are in Categories 2, 3, or 4. Table A.3 shows Officer AFSCs in all categories.

Table A.1. Enlisted Air Force Specialty Codes in Category 1 with Program Guidance Letter Change Data and Statistics

AFSC	AFSC Description	Initial PGL (PGL Change) by Year								Avg Initial PGL	Avg % Change	% of Years >%25 Change	Category
		2009	2010	2011	2012	2013	2014	2015	2016				
3P031	Security Forces	4741(470)	4689(18)	4812(-332)	4605(-197)	4571(-309)	4655(-280)	4333(-36)	4333(169)	4592	-3%	0%	1
4N031	Aerospace Medical Service	1349(257)	1405(-50)	1198(33)	1310(41)	1281(2)	1322(-37)	5570(-5)	5571(-1)	2376	1%	0%	1
3E731	Fire Protection	1346(135)	1363(-114)	1496(-99)	1329(-29)	1310(-87)	1256(-37)	1158(25)	1143(29)	1300	-5%	0%	1
2T231	Air Trans	1282(26)	1055(1)	965(16)	1174(29)	1210(-32)	1325(-51)	1272(-70)	1135(61)	1177	0%	0%	1
1N431	Fusion Analyst	1171(189)	1099(7)	1054(116)	1304(144)	1097(0)	1225(153)	817(221)	1073(195)	1105	12%	14%	1
2W031	Munitions Sys	878(157)	970(-48)	1069(-13)	1009(-42)	1085(4)	1209(-126)	1058(-195)	930(14)	1026	-5%	0%	1
2S031	Materiel Mgt	1171(-277)	1159(-197)	1146(-287)	1004(-81)	966(-45)	857(19)	830(45)	956(93)	1011	-3%	14%	1
3M031	Services	1127(180)	1057(-114)	887(-99)	810(-14)	837(-41)	825(-30)	799(93)	871(44)	902	-2%	0%	1
3D031	Knowledge Operations Management		1058(-133)	1009(-109)	905(-26)	877(-5)	874(-123)	827(-492)	292(40)	835	-12%	14%	1
3S031	Personnel	779(97)	743(-102)	776(17)	834(-40)	794(-15)	783(31)	806(115)	1152(-119)	833	0%	0%	1
3E431	Water and Fuel Systems Maintenance	911(120)	310(0)	863(25)	870(15)	828(-83)	807(-18)	774(-71)	799(31)	770	0%	0%	1
3D032	Cyber Systems Operations		700(57)	710(-14)	732(20)	761(-28)	834(-45)	839(-162)	729(6)	758	-3%	0%	1
3D132	Cyber Transport Systems		662(-42)	688(-110)	680(-11)	690(25)	720(-22)	733(-45)	761(21)	705	-2%	0%	1
1C131	Air Traffic Control	666(0)	716(0)	671(0)	708(-20)	710(-95)	633(-109)	597(-101)	515(-4)	652	-7%	0%	1
2A636	Acft Elect & Envir Systems	554(168)	561(-17)	614(-4)	713(4)	738(-4)	714(-122)	655(-42)	593(1)	643	-5%	0%	1
1N031	Operations Intelligence	663(-191)	668(-75)	777(-97)	642(0)	698(-41)	695(-135)	482(44)	486(73)	639	-2%	0%	1
3D133	RF Transmission Systems		708(-181)	624(-56)	700(-89)	690(-33)	630(-30)	596(-108)	497(-1)	635	-10%	0%	1
8F000	First Sergeant	684(0)	610(0)	640(0)	625(0)	625(0)	625(0)	625(-9)	616(30)	631	0%	0%	1
1N131A	Geospatial Intelligence - Analyst		802(-50)	829(-136)	643(27)	663(-38)	595(-46)	473(-64)	400(-41)	629	-8%	0%	1
1W031	Weather	683(-27)	642(-27)	624(10)	655(-43)	598(5)	579(-37)	616(-78)	577(-7)	622	-4%	0%	1
2A632	Aerospace Ground Equip	591(195)	573(-27)	592(-9)	740(-40)	713(-77)	618(-63)	580(-65)	553(31)	620	-5%	0%	1
3D131	Client Systems		644(50)	594(-38)	621(-99)	566(-18)	555(12)	541(-25)	679(93)	600	-2%	0%	1
3E031	Electrical Systems	823(88)	732(-19)	605(-60)	567(5)	426(-98)	463(-12)	443(-28)	433(-25)	562	-4%	0%	1
3P031A	Security Forces - Mil Working Dog Handler	532(48)	513(-5)	498(-6)	516(3)	554(-23)	527(-115)	508(-44)	478(26)	516	-4%	0%	1
4A031	Health Services Management	566(158)	440(0)	415(-55)	455(-30)	398(10)	427(-5)	684(-2)	680(0)	508	-3%	0%	1
2F031	Fuels	496(187)	494(-27)	483(-26)	582(-58)	534(-38)	557(-118)	442(11)	473(35)	508	-6%	0%	1
1A231	Aircraft Loadmaster	443(64)	485(-41)	538(8)	581(121)	634(22)	374(-119)	383(65)	346(3)	474	2%	0%	1
3E131	Heating, Ventilation, AC, & Refrigeration	727(100)	317(-2)	554(-28)	548(-33)	381(-97)	418(-4)	396(-44)	418(17)	470	-6%	0%	1
2A733	Aircraft Structural Maintenance	482(48)	508(-42)	545(-63)	517(-48)	451(19)	409(-11)	412(-46)	398(68)	465	-3%	0%	1
3E231	Pavements and Construction Equipment	360(60)	520(8)	498(0)	543(0)	532(-25)	451(-9)	457(-46)	346(16)	463	-2%	0%	1
2A634	Aircraft Fuel Systems	358(92)	373(-3)	419(-14)	470(-5)	462(40)	475(-91)	452(-80)	365(33)	422	-3%	0%	1
3E331	Structural	323(64)	485(7)	475(0)	473(-28)	468(-60)	375(-25)	355(-31)	311(28)	408	-5%	0%	1
2A635	Aircraft Hydraulic Systems	356(88)	387(-21)	392(-15)	414(10)	422(2)	458(-68)	394(2)	373(-11)	400	-3%	0%	1
1P031	Aircrew Flight Equipment			465(-72)	425(-30)	404(-15)	362(-11)	366(-34)	365(66)	398	-2%	0%	1
2W131F	Acft Armament Systems - F-16	411(109)	354(-45)	495(31)	474(-27)	451(-17)	350(-25)	350(-61)	285(-1)	396	-6%	0%	1
6F031	Financial Management & Comptroller	406(131)	416(15)	418(-7)	384(-8)	377(-36)	342(-28)	309(7)	312(7)	371	-1%	0%	1
2T131	Vehicle Ops	426(87)	378(-49)	343(-59)	355(-10)	381(-48)	343(-25)	320(10)	390(44)	367	-4%	0%	1
2T331	Vehicle & Vehicular Equip Maint	377(33)	329(-15)	318(6)	389(-21)	367(-11)	370(-14)	367(-28)	341(6)	357	-3%	0%	1
3D033	Cyber Surety		302(-30)	300(-2)	378(-30)	352(4)	371(-2)	353(-4)	337(21)	342	-1%	0%	1
1C331	Command Post	264(8)	316(6)	385(3)	395(3)	358(-13)	315(-6)	290(-72)	358(-11)	335	-5%	0%	1
3S231	Education and Training	402(0)	335(0)	312(-50)	294(0)	325(0)	338(5)	313(-74)	304(29)	328	-5%	0%	1
3P031B	Security Forces - Combat Arms	375(134)	337(0)	315(0)	320(-1)	320(0)	320(-35)	320(-25)	295(0)	325	-3%	0%	1
4A231	Biomedical Equipment	326(46)	278(-1)	374(-8)	365(5)	371(-1)	368(4)	357(-31)	128(0)	321	-1%	0%	1
1C032	Aviation Resource Mgt	269(83)	214(52)	339(-66)	325(-2)	369(-3)	392(-32)	336(-31)	318(21)	320	-2%	0%	1
3E032	Electrical Power Production	343(94)	330(4)	337(-10)	320(6)	304(-32)	315(-17)	299(-16)	255(65)	313	0%	0%	1
2A631C	Aerospace Propulsion - TF33, CF6, F103, F108, F117, TFE-731, TF34, TF39, PW 2040, F138 Jet Engines	303(81)	279(-28)	305(-20)	325(2)	286(0)	260(2)	308(-52)	270(46)	292	-2%	0%	1
4Y031	Dental Assistant	312(108)	275(0)	248(-47)	283(-12)	261(-6)	258(-11)	252(-5)	352(-78)	280	-8%	0%	1
1C431	Tac Air Con Party (TACP)	237(31)	221(8)	273(17)	313(3)	343(-17)	289(-12)	304(-42)	219(10)	275	-1%	0%	1
2T031	Traffic Mgt	307(71)	293(-13)	281(-51)	241(20)	273(-4)	257(-5)	247(35)	286(71)	273	3%	0%	1
4A131	Medical Materiel	238(52)	216(-16)	182(-34)	178(7)	175(-3)	175(-7)	469(0)	444(11)	260	-4%	0%	1
1C531	Command & Control Battle Mgt Ops	226(119)	229(13)	263(-15)	234(5)	275(-21)	257(-1)	199(-25)	271(-55)	244	-5%	0%	1
2A732	Nondestructive Inspection	221(38)	228(0)	231(-34)	240(-15)	210(28)	216(-28)	183(0)	211(24)	218	-1%	0%	1
2A631E	Aerospace Propulsion - F101, F110, F118 Jet Eng	184(66)	163(12)	185(-7)	182(47)	247(14)	240(-15)	260(-34)	269(-19)	216	2%	14%	1
2W131E	Acft Armament Systems - F-15	218(67)	200(-26)	178(-24)	175(-25)	164(74)	236(-12)	262(2)	258(-40)	211	-2%	14%	1
4C031	Mental Health Service	173(23)	166(-5)	100(6)	107(16)	103(5)	115(-7)	457(13)	458(0)	210	3%	0%	1
2A731	Aircraft Metals Technology	160(35)	187(-41)	195(5)	253(-16)	218(25)	245(-56)	229(-34)	177(12)	208	-7%	0%	1
2G031	Logistics Plans	225(30)	210(-21)	202(-32)	181(-18)	194(-8)	193(-19)	198(13)	175(0)	197	-6%	0%	1
3E931	Emergency Management	238(37)	165(-9)	205(-8)	197(-10)	192(-4)	194(-15)	181(16)	200(-4)	197	-2%	0%	1
3E531	Engineering	183(21)	204(-6)	211(-22)	215(-1)	174(19)	196(-6)	184(-5)	195(14)	195	0%	0%	1

AFSC	AFSC Description	Initial PGL (PGL Change) by Year								Avg Initial PGL	Avg % Change	% of Years >%25 Change	Category
		2009	2010	2011	2012	2013	2014	2015	2016				
4P031	Pharmacy	128(41)	112(4)	97(16)	142(2)	134(-5)	127(-5)	376(-15)	404(-32)	190	0%	0%	1
2R131	Maint Mgt Production	206(86)	178(-21)	192(4)	177(4)	214(2)	193(-4)	178(6)	160(7)	187	0%	0%	1
2A631D	Aerospace Propulsion - F100, F119, F135 Jet Engines	208(73)	195(-37)	193(-25)	170(13)	195(-20)	173(-21)	153(-18)	153(15)	180	-7%	0%	1
4R031	Diagnostic Imaging	121(25)	106(5)	110(-4)	113(5)	107(-8)	107(-14)	396(-15)	380(16)	180	-2%	0%	1
3E831	Explosive Ordnance Disposal	204(-1)	249(14)	156(9)	158(-6)	125(2)	117(16)	165(12)		171	4%	0%	1
3E631	Operations Management	119(54)	122(8)	137(51)	190(42)	219(-1)	212(-15)	197(-16)	156(23)	169	9%	14%	1
1N231C	Signals Intel Analyst - Communications				180(-11)	174(-25)	144(8)	154(-1)	189(152)	168	13%	20%	1
4B031	Bioenvironmental Engineering	197(53)	181(27)	142(3)	157(11)	154(-4)	158(-14)	149(0)	143(4)	160	2%	0%	1
4T031	Medical Laboratory	163(57)	153(0)	149(3)	180(-7)	168(-1)	169(-17)	151(-28)	146(28)	160	-2%	0%	1
2A735	Low Observable Aircraft Structural Maintenance		118(6)	222(-25)	158(23)	136(44)	180(-43)	172(17)	125(9)	159	5%	14%	1
5J031	Paralegal	172(32)	155(2)	159(0)	140(19)	144(15)	164(-29)	159(-64)	140(8)	154	-4%	14%	1
1A331	Airborne Mission Systems Operator	144(46)	178(-27)	152(-11)	158(-9)	153(-23)	121(-10)	130(39)	186(-2)	153	-3%	14%	1
1C731	Airfield Mgt	160(46)	163(-1)	156(-8)	151(4)	158(-19)	153(-12)		134(-7)	153	-2%	0%	1
3D136	Airfield Systems		143(10)	126(6)	164(-35)	147(0)	139(-10)	131(-131)		142	-20%	17%	1
2R031	Maint Mgt Analysis	149(92)	136(-5)	138(2)	141(6)	165(-5)	142(-9)	130(-2)	130(3)	141	-1%	0%	1
7S031	Special Investigations	129(0)	105(0)	150(0)	150(0)	150(-2)	141(-6)	148(-25)	123(0)	138	-3%	0%	1
1T231	Pararescue	191(42)	165(0)	101(1)	136(0)	133(33)	128(-5)	114(6)	123(0)	136	4%	0%	1
2P031	Precision Measurement Equip Lab	133(24)	141(-12)	148(-31)	137(-27)	120(-3)	149(-13)	124(-3)	118(-8)	134	-10%	0%	1
8M000	Postal	166(0)	125(0)	110(0)	122(5)	125(1)	127(-7)			129	0%	0%	1
4N131	Surgical Service	84(22)	75(0)	63(13)	81(7)	86(-3)	94(-12)	275(-6)	273(11)	129	2%	0%	1
3S331	Equal Opportunity	143(0)	119(0)	124(0)	127(5)	129(-7)				128	0%	0%	1
3D135	Ground Radar Systems		153(0)	123(-9)	128(5)	141(-5)	144(-5)	78(-78)		128	-13%	17%	1
2W131C	Acft Armament Systems - A-10	103(20)	92(0)	120(5)	130(-1)	176(-10)	136(-23)	118(-25)	147(7)	128	-5%	0%	1
1C231	Combat Control	129(106)	128(-2)	130(-4)	122(6)	128(8)	133(-22)	137(-18)	119(5)	127	-1%	0%	1
8B100	Military Training Leader	112(20)	137(25)	112(5)	112(-13)	112(1)	111(12)	125(-2)	178(-56)	125	-1%	14%	1
1S031	Safety	120(0)	117(0)	110(1)	124(19)	132(12)	130(-13)	134(-15)	101(7)	121	2%	0%	1
4N051C	Aerospace Medical Service - IDMT	96(0)	96(0)	130(-28)	132(0)	134(0)	131(-20)	111(-39)	87(0)	115	-10%	14%	1
1N131B	Geospatial Intelligence - Targeteer		142(142)	150(-5)	131(-4)	120(-26)	70(16)	70(59)	86(13)	110	16%	17%	1
3D134	Spectrum Operations		132(0)	136(7)	107(8)	111(0)	91(-8)	98(-14)	85(0)	109	-1%	0%	1
4Y032	Dental Laboratory	126(20)	117(-10)	123(-2)	122(2)	104(-2)	102(-9)	110(0)	62(2)	108	-2%	0%	1
2W231	Nuclear Weapons	88(46)	143(14)	143(-8)	131(-16)	98(-4)	96(-21)	83(2)	78(21)	108	-1%	14%	1
1N231A	Signals Intel Analyst - Electronic				98(23)	102(5)	98(-12)	87(9)	117(68)	100	17%	20%	1
2T337	Veh Mgt & Analysis	113(49)	120(-5)	106(-14)	114(-8)	102(-2)	83(-5)	80(9)	78(4)	100	-2%	0%	1
1T031	Survival, Evasion, Resist, & Escape	122(14)	119(2)	114(0)	136(1)	70(-1)	44(0)	55(0)	50(1)	89	0%	0%	1
2A532B	Helicopter/ Tiltrotor Aircraft Maintenance - H-60	65(20)	71(0)	84(2)	109(1)	113(-16)	100(-22)	78(-18)	70(5)	86	-7%	0%	1
2W131N	Acft Armament Systems - F-22	107(30)	94(-17)	93(-11)	95(-1)	93(0)	78(-12)	68(3)	61(18)	86	-1%	14%	1
2W131K	Acft Armament Systems - B-52/B-2	53(15)	74(-9)	76(15)	78(33)	108(-6)	104(-1)	95(-1)	90(3)	85	7%	14%	1
3N032	Broadcast Journalist	59(6)	92(-18)	104(1)	80(8)	85(-4)	85(-4)	83(-6)	78(2)	83	-3%	0%	1
5R031	Chaplain Assistant	67(-50)	75(0)	78(1)	71(4)	87(0)	82(7)	82(-11)	95(0)	80	0%	0%	1
4H031	Cardiopulmonary Laboratory	68(11)	69(-6)	77(1)	70(8)	80(-1)	84(-9)	77(-2)	73(-1)	76	-2%	0%	1
4V031	Optometry	61(4)	55(4)	58(5)	75(2)	72(-3)	71(2)	118(0)	91(4)	75	3%	0%	1
4D031	Diet Therapy	61(37)	45(-2)	35(-1)	44(-1)	43(-4)	40(-4)	151(-2)	166(-18)	73	-6%	0%	1
3E433	Pest Management	76(11)	73(0)	72(3)	93(7)	64(-25)	74(-1)	66(-12)	50(8)	71	-4%	14%	1
4J032	Physical Medicine	48(28)	35(0)	31(-2)	28(9)	33(-5)	29(-2)	182(-7)	179(6)	71	0%	14%	1
3D034	Computer Systems Programming		68(-5)	78(1)	94(-21)	70(-5)	57(-2)	65(-6)	60(1)	70	-7%	0%	1
2T332C	Spec Veh Maint - MHE	54(19)	74(28)	83(-18)	91(-12)	72(-4)	57(-5)	54(-4)	#DIV/0!	69	0%	17%	1
2M032	Msl & Space Sys Maint	54(16)	66(10)	68(1)	65(1)	66(-8)	59(0)	59(-5)	70(-4)	63	-1%	0%	1
9S100	Scientific Applications Specialist	61(10)	70(-8)	59(8)	53(3)	63(0)	63(1)	65(-4)	61(7)	62	3%	0%	1
2T332A	Spec Veh Maint - Fire Fighting Vehicles	58(18)	53(-4)	57(3)	70(-2)	60(2)	55(-2)	53(1)	#DIV/0!	58	-1%	0%	1
2M033	Msl & Space Facilities	44(-5)	47(0)	43(0)	51(-1)	50(-2)	52(-4)	48(-2)	85(1)	53	-2%	0%	1
2W131L	Acft Armament Systems - B-1	50(15)	45(-11)	42(-10)	39(0)	55(-2)	58(-14)	47(0)	47(12)	48	-7%	14%	1
2A532D	Helicopter/ Tiltrotor Aircraft Maintenance - CV-22	10(4)	46(16)	43(7)	56(0)	51(2)	57(-9)	53(13)	65(11)	48	14%	14%	1
2W131Z	Acft Armament Systems - Other	35(12)	25(-3)	37(7)	51(3)	80(5)	37(-6)	33(-5)	30(22)	41	9%	14%	1
2M031B	Msl & Space Sys Elect Maint - ALCM	22(1)	54(-16)	54(1)	48(7)	45(-5)	34(0)	34(2)	26(5)	40	0%	14%	1
4N131C	Surgical Service - Orthopedics	20(2)	25(0)	33(0)	20(0)	15(0)	15(-4)	91(-7)	96(0)	39	-5%	14%	1
4M031	Aerospace and Operational Physiology	39(12)	29(-2)	31(-1)	36(1)	34(-4)	40(-8)	31(-4)	31(4)	34	-4%	0%	1
4R031B	Diagnostic Imaging - Ultrasound	23(0)	23(0)	23(0)	23(0)	23(0)	23(0)	23(0)	23(0)	23	0%	0%	1
4N131D	Surgical Service - Otolaryngology	4(0)	5(0)	5(0)	5(0)	5(0)	5(0)	30(-1)	30(0)	11	0%	0%	1
4R031A	Diagnostic Imaging - Nuclear Medicine	8(0)	10(0)	8(0)	8(0)	8(0)	8(3)	17(-3)	14(0)	10	3%	14%	1
4T032	Histopathology	15(8)	13(0)	11(-1)	14(0)	7(0)	7(0)	7(0)	7(3)	10	5%	14%	1
3N231	Premier Band	9(-9)	11(-1)	10(0)	12(0)	9(-2)	9(-1)	9(0)	8(4)	10	1%	14%	1
4N131B	Surgical Service - Urology	3(2)	5(0)	5(4)	5(0)	4(0)	4(0)	15(0)	15(0)	7	11%	14%	1
2W131	Acft Armament Systems	5(0)	5(0)	5(0)	5(0)	5(0)	5(0)	0(0)	0(0)	4	0%	0%	1

Table A.2. Enlisted Air Force Specialty Codes in Categories 2, 3, or 4 with Program Guidance Letter Change Data and Statistics

AFSC	AFSC Description	Initial PGL (PGL Change) by Year								Avg Initial PGL	Avg % Change	% of Years >%25	Category
		2009	2010	2011	2012	2013	2014	2015	2016				
1N331	Cryptologic Language Analyst		1299(106)	1374(100)	1452(337)	1936(-24)	338(-137)	258(-138)	112(-60)	967	-10%	29%	2
8R000	Enlisted Accessions Recruiter	674(172)	573(-12)	534(-192)	572(-237)	576(-3)	574(6)	574(6)	514(-26)	574	-12%	29%	2
6C031	Contracting	278(90)	419(140)	474(2)	472(6)	386(-7)	405(-115)	312(-115)	158(28)	363	-1%	43%	2
1N231	Signals Intel Analyst	194(61)	185(0)	185(-28)	#DIV/0!			15(381)	832(-224)	282	-14%	33%	2
1N431B	Analysis and Production					354(-115)	255(-25)	238(-8)	232(103)	270	0%	50%	2
1A131	Flight Engineer	339(0)	315(0)	314(0)	329(-24)	327(-65)	181(-113)	202(-89)	129(62)	267	-11%	43%	2
3D137	Cable and Antenna Systems		245(39)	266(-19)	349(-138)	224(-109)	242(-11)	275(-43)	252(-30)	265	-13%	29%	2
1U031	Remotely Piloted Aircraft (RPA) Sensor Operator		216(286)	353(25)	329(19)	202(121)	314(-120)	323(-171)	79(174)	259	17%	67%	2
8B000	Military Training Instructor	234(40)	192(25)	175(0)	175(8)	175(47)	176(88)	264(-62)	224(10)	202	11%	29%	2
8T000	Professional Military Education Instructor	140(0)	140(0)	150(71)	214(0)	214(0)	214(0)	214(-64)	150(0)	180	2%	29%	2
1A031	In-Flt Refueling	174(61)	173(2)	188(12)	181(52)	254(-74)	143(0)	147(16)	144(48)	176	7%	43%	2
1C631	Space Sys Ops	127(12)	140(-10)	164(26)	233(-35)	236(-47)	188(-19)	156(-21)	132(12)	172	-10%	29%	2
4E031	Public Health	164(41)	158(-10)	138(59)	160(65)	196(6)	205(-26)	174(-1)	167(4)	170	10%	29%	2
2A633	Aircrew Egress Systems	122(47)	136(-5)	181(-27)	182(-24)	192(14)	169(-18)	187(-52)	137(37)	163	-6%	29%	2
1A832	Airborne Intelligence, Surveillance, & Reconnaissance (ISR) Operator		80(0)	90(10)	85(56)	112(18)	112(-69)	75(-31)	84(-9)	91	-3%	43%	2
8G000	Honor Guard	84(30)	97(2)	95(-27)	83(-5)	67(5)	74(-2)	72(-7)	72(24)	81	-1%	29%	2
2A031P	Avionics Sensor Systems and Electronic Warfare Sys		90(-52)	84(-12)	87(5)	66(13)	79(-25)	54(-1)	52(0)	73	-9%	29%	2
3S131	Equal Opportunity	47(0)	82(35)	57(8)	60(2)	60(21)	71(-17)	71(-16)	61(5)	64	13%	29%	2
2M031A	Msl & Space Sys Elect Maint - ICBM	45(5)	72(47)	72(2)	71(5)	65(-17)	60(-10)	50(-3)	48(7)	60	13%	29%	2
2A333H	Tactical Aircraft Maintenance	35(0)	28(-8)	44(66)	84(-84)	74(-74)				53	-19%	100%	2
1A631	Flight Attendant	53(0)	53(0)	54(0)	59(-8)	49(-13)	17(-30)	47(-18)	22(15)	44	-11%	57%	2
3N131	Regional Band	32(28)	43(-18)	44(-4)	45(0)	41(-8)	44(-20)	28(-2)	20(15)	37	-7%	43%	2
1W032	Special Operations Weather	12(106)	32(12)	32(0)	27(8)	32(2)	33(-6)	30(-2)	41(-2)	30	9%	29%	2
4N031B	Aerospace Medical Service - Neurodiagnostic Medical Technician	6(0)	3(0)	3(0)	3(0)	3(0)	3(3)	6(0)	10(-4)	5	10%	29%	2
4J032A	Physical Medicine - Orthotic	3(0)	3(0)	3(-1)	3(0)	3(-1)	2(0)	2(0)	0(2)	2	-11%	33%	2
2A333E	Tactical Aircraft Maintenance - A-10/U-2	133(31)	105(10)	186(34)	217(81)	247(49)	204(-48)	157(-66)	44(69)	162	25%	43%	3
2A334C	F-16 Avionics				0(140)	33(146)	197(-53)	180(-42)	206(-28)	123	94%	50%	3
1B431	Cyber Warfare Ops		91(0)	53(21)	38(34)	65(49)	102(31)	132(67)	189(27)	96	43%	71%	3
2A832C	Mobility Air Forces Integrated Instrument & Flt Control Sys				0(33)	5(61)	96(-14)	82(-14)	75(-6)	52	295%	25%	3
2A831C	Mobility Air Forces Integrated Comm/Nav/ Mission Sys				5(42)	0(62)	71(-15)	66(12)	76(8)	44	212%	25%	3
2A533A	Mobility Air Forces Electronic Warfare Sys	432(126)	444(-14)	454(-33)	485(-35)	409(-409)				445	-29%	25%	4
2A533B	Mobility Air Forces Electronic Warfare Sys	423(131)	421(-7)	425(14)	510(-24)	391(-391)				434	-26%	25%	4
2A631B	Aerospace Propulsion	242(113)	245(-27)	249(-27)	265(-4)	305(-305)				261	-31%	25%	4
2A533C	Mobility Air Forces Electronic Warfare Sys	197(81)	206(-5)	207(4)	267(-20)	221(-221)				220	-27%	25%	4
2A333K	Tactical Aircraft Maintenance	145(39)	209(-19)	149(-11)	183(-183)	200(-200)				177	-54%	50%	4
2A031S	Avionics Test Stn & Cmpnts		126(-50)	113(-13)	130(-6)	109(14)	98(-98)			115	-26%	40%	4
1A431	Airborne Operations	123(88)	156(-18)	122(-25)	104(-7)	100(-14)	75(-12)	75(-75)		108	-28%	17%	4
3N031	Public Affairs	97(4)	88(11)	61(0)	84(0)	88(-88)				84	-21%	25%	4
3N034	Still Photography	63(8)	101(1)	95(-1)	78(0)	74(-74)				82	-25%	25%	4
1A731	Aerial Gunner	54(17)	68(0)	64(-1)	65(-13)	66(-66)				63	-30%	25%	4
2A533D	Mobility Air Forces Electronic Warfare Sys	37(25)	27(-19)	33(1)	38(1)	33(-33)				34	-34%	50%	4

NOTE: Some AFSCs have missing data for some years. Given that the source of these data is the PGLs, it is likely that there were no requirements for these cases in these years. If resourcing decisions are to be made specific to these cases, confirming that the historical PGL records have no requirements for these years is prudent.

Table A.3. Officer Air Force Specialty Codes in All Categories with Program Guidance Letter Change Data and Statistics

AFSC	AFSC Description	Initial PGL (PGL Change) by Year								Avg Initial PGL	Avg % Change	% of Years >%25	Category
		2010	2011	2012	2013	2014	2015	2016	2017				
14N1	Intelligence	490(0)	589(-47)	572(66)	557(-25)		450(62)	476(34)	525(-25)	523	0%	0%	1
17D1	Network Operations	0(118)	321(22)	316(39)	360(6)		380(17)	356(59)	474(-23)	315	2%	14%	1
21A1	Aircraft Maintenance	309(-11)	277(-16)	248(-2)	274(29)		298(32)	292(31)	326(-4)	289	1%	0%	1
21R1	Logistics Readiness	242(-4)	257(-25)	212(23)	202(8)		231(31)	233(17)	225(-5)	229	2%	0%	1
64P1	Contracting	293(-2)	294(-7)	269(3)	-14(14)		126(2)	124(15)	146(-2)	177	3%	14%	1
65F1	Financial Management	134(-4)	140(0)	141(-8)	124(9)		110(8)	111(18)	135(-1)	128	3%	0%	1
32E1X	Civil Engineering	76(24)	102(1)	97(8)	101(2)		87(7)	80(9)	92(0)	91	8%	0%	1
31P1	Security Forces	115(0)	91(-4)	75(9)	83(3)		75(11)	80(10)	90(-2)	87	4%	0%	1
35P1	Public Affairs	103(-30)	91(-6)	64(3)	61(-4)		69(-12)	46(0)	61(-7)	71	-11%	14%	1
13M1	Airfield Operation	68(0)	49(-3)	49(4)	67(1)		70(0)	63(3)	69(1)	62	1%	0%	1
71S1	Special Investigations		0(22)	91(8)	67(-4)		73(-11)	63(3)	64(-3)	60	-3%	17%	1
15W1	Weather	55(-1)	52(-2)	45(8)	53(3)		51(-3)	59(0)	65(-2)	54	1%	0%	1
13L1	Air Liaison Officer	20(0)	42(-1)	64(4)	58(-1)	Data Not Obtained	36(9)	64(6)	52(-1)	48	4%	0%	1
21M1A	Munitions and Missile Maintenance	28(8)	19(1)	31(0)	34(1)		49(0)	37(2)	35(0)	33	5%	0%	1
21M1N	Munitions and Missile Maintenance	17(0)	21(-1)	18(-1)	19(0)		23(0)	25(2)	21(0)	21	-1%	0%	1
13D1	Combat Rescue Officer				21(1)		24(1)	18(-1)	16(2)	20	3%	0%	1
13N1	Nuclear and Missile Operations				180(-12)		180(-59)	156(17)	161(-34)	169	-18%	50%	2
13S1A	Space Operations	56(-4)	51(-12)	34(0)	20(29)		74(4)			47	2%	40%	2
61A1	Operations Research Analyst	27(26)	53(0)	45(-10)	43(-6)		40(-1)	36(1)	36(-9)	40	4%	43%	2
13S1E	Space Operations	30(0)	29(-9)	23(-2)	11(26)		67(10)			32	1%	40%	2
13S1B	Space Operations	18(-4)	13(-2)	10(1)	18(0)		16(18)			15	4%	40%	2
32E3G	Civil Engineering	23(0)	13(0)	12(7)	12(0)		10(1)	16(0)	11(4)	14	12%	29%	2
13C1	Special Tactics				14(0)		13(1)	13(5)	12(5)	13	16%	50%	2
21M1I	Munitions and Missile Maintenance		14(1)	10(-1)	12(0)		9(6)	12(5)	14(1)	12	16%	40%	2
13S1C	Space Operations	245(-74)	187(-15)	176(13)	7(-7)					154	-33%	50%	4
13S1D	Space Operations	3(0)	2(-1)	8(1)	6(4)		19(-7)			8	-24%	40%	4

Appendix B. Officer Air Force Specialty Code Program Guidance Letter Changes

Figure B.1 shows the percentage of AFSCs that had a percentage increase of 5, 15, 25, and greater than 35 percent. Unlike the enlisted AFSCs in 2016 as shown in Figure 9.3, officer AFSCs experienced only small PGL increases. This is seen by the greater-than-15-percent change remaining relatively constant in the recent four years. The 2016 ">5% increase" is large in 2016, but it is not without precedent, given that 2015 and 2013 were similar. The large difference between the ">5% increase" and the ">15% increase" lines (i.e., the blue and red lines, respectively) indicates that, for most AFSCs, the changes were between 5 and 15 percent.

Figure B.1. Percentage of Air Force Specialty Codes with Program Guidance Letter Changes

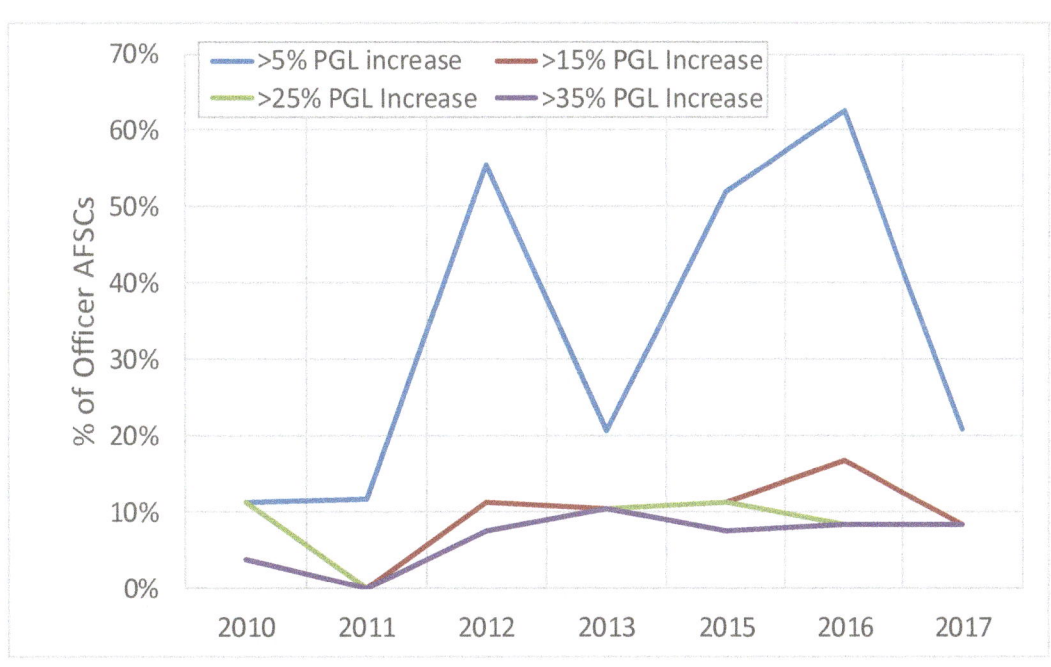

References

Aghina, Wouter, Aaron De Smet, and Kirsten Weerda, "Agility: It Rhymes with Stability," *McKinsey Quarterly*, December 2015. As of October 25, 2017: https://www.mckinsey.com/business-functions/organization/our-insights/agility-it-rhymes-with-stability

Air Education and Training Command, *On Learning: The Future of Air Force Education and Training*, Randolph Air Force Base, Tex., January 30, 2008.

Air Education and Training Command, "Continuum of Learning: A Fundamentally Different Approach to Airmen's Development," December 19, 2017. As of November 11, 2018: https://www.airman.af.mil/Portals/17/001%20Home%20Page/DAY%202%20-%202%20-%20CoL%20Engagement%20Standard%20Briefing_PACE%20Summit_v3.pdf?ver=2017-12-19-172816-107

Air Force Instruction 36-2101, *Classifying Military Personnel (Officer and Enlisted)*, Washington, D.C.: Secretary of the Air Force, March 9, 2017.

Allen, I. Elaine, and Jeff Seaman, *Going the Distance: Online Education in the United States*, San Francisco: Babson Survey Research Group and Quahog Research Group, LLC, 2011.

American Association of Community Colleges, *2017 Fact Sheet*, Washington, D.C., June 2017. As of April 21, 2020: https://www.aacc.nche.edu/wp-content/uploads/2017/09/AACCFactSheet2017.pdf

Azadegan, Arash, Pankaj C. Patel, Abouzar Zangoueinezhad, and Kevin Linderman, "The Effect of Environmental Complexity and Environmental Dynamism on Lean Practices," *Journal of Operations Management*, Vol. 31, No. 4, May 2013, pp. 193–212.

Bettinger, Eric P., and Bridget Terry Long, "Does Cheaper Mean Better? The Impact of Using Adjunct Instructors on Student Outcomes," *Review of Economics and Statistics*, Vol. 92, No. 3, August 2010, pp. 598–613.

Bhattacharya, Arijit, John Geraghty, Paul Young, and P. J. Byrne, "Design of a Resilient Shock Absorber for Disrupted Supply Chain Networks: A Shock-Dampening Fortification Framework for Mitigating Excursion Events," *Production Planning and Control*, Vol. 24, No. 8–9, September 2013, pp. 721–742.

Brigham, Cathy, and Rebecca Klein-Collins, *Availability, Use and Value of Prior Learning Assessment Within Community Colleges*, Chicago: Council for Adult and Experiential Learning, 2010. As of April 21, 2020: https://files.eric.ed.gov/fulltext/ED524751.pdf

Bushman, Mary F., and John E. Dean, "Outsourcing of Non-Mission-Critical Functions: A Solution to the Rising Cost of College Attendance," *Course Corrections: Experts Offer Solutions to the College Cost Crisis*, Indianapolis, Ind.: Lumina Foundation for Education, October 2005, pp. 6–19.

Carnegie Foundation for the Advancement of Teaching, *50-State Scan of Course Credit Policies*, working draft, Menlo Park, Calif.: Carnegie Foundation for the Advancement of Teaching, May 2014.

Center for Digital Education, *Realizing the Full Potential of Blended Learning*, Folsom, Calif., 2012.

Chopra, Sunil, Gilles Reinhardt, and Usha Mohan, "The Importance of Decoupling Recurrent and Disruption Risks in a Supply Chain," *Naval Research Logistics*, Vol. 54, No. 5, August 2007, pp. 544–555.

Christofferson, Jay P., Kristen Wynn, and Jay S. Newitt, "Assessing Construction Management Higher Education Strategies: Increasing Demand, Limited Resources, and Over-Enrollment," *International Journal of Construction Education and Research*, Vol. 2, No. 3, 2006, pp. 181–192.

Christopher, Martin, "The Agile Supply Chain: Competing in Volatile Markets," *Industrial Marketing Management*, Vol. 29, No. 1, January 2000, pp. 37–44.

Christopher, Martin, and Hau Lee, "Mitigating Supply Chain Risk Through Improved Confidence," *International Journal of Physical Distribution and Logistics Management*, Vol. 34, No. 5, June 2004, pp. 388–396.

Christopher, Martin, and Denis Towill, "An Integrated Model for the Design of Agile Supply Chains," *International Journal of Physical Distribution and Logistics Management*, Vol. 31, No. 4, May 2001, pp. 235–246.

Clark, Ruth C., and Richard E. Mayer, *E-Learning and the Science of Instruction: Proven Guidelines for Consumers and Designers of Multimedia Learning*, 3rd ed., San Francisco: Pfeiffer, 2011.

Constant, Louay, Shelly Culbertson, Cathy Stasz, and Geroges Vernez, *Improving Technical Vocational Education and Training in the Kurdistan Region—Iraq*, Santa Monica, Calif.: RAND Corporation, RR-277-KRG, 2014. As of September 29, 2017: https://www.rand.org/pubs/research_reports/RR277.html

Craighead, Christopher W., Jennifer Blackhurst, M. Johnny Rungtusanatham, and Robert B. Handfield, "The Severity of Supply Chain Disruptions: Design Characteristics and Mitigation Capabilities," *Decision Sciences*, Vol. 38, No. 1, February 2007, pp. 131–156.

Cull, Selby, Don Reed, and Karin Kirk, "Student Motivation and Engagement in Online Courses," *Teaching Geoscience Online—A Workshop for Digital Faculty*, 2010. As of September 29, 2017:
http://serc.carleton.edu/NAGTWorkshops/online/motivation.html

Daugherty, Lindsay, Van L. Davis, and Trey Miller, *Competency-Based Education Programs in Texas: An Innovative Approach to Higher Education*, Santa Monica, Calif.: RAND Corporation, RR-1239-1-CFAT, 2015. As of October 25, 2018:
https://www.rand.org/pubs/research_reports/RR1239-1.html

Desrochers, Donna M., and Richard L. Staisloff, *Competency-Based Education: A Study of Four New Models and Their Implications for Bending the Higher Education Cost Curve*, Indianapolis, Ind.: Lumina Foundation, October 11, 2016. As of October 25, 2018:
http://rpkgroup.com/wp-content/uploads/2016/10/rpkgroup_cbe_business_model_report_20161018.pdf

Devaux, Axelle, Julie Belanger, Sarah Grand-Clement, and Catriona Manville, *Education: Digital Technology's Role in Enabling Skills Development for a Connected World*, Santa Monica, Calif.: RAND Corporation, PE-238-CI, 2017. As of September 29, 2017:
https://www.rand.org/pubs/perspectives/PE238.html

Durach, Christian F., Andreas Wieland, and José A. D. Machuca, "Antecedents and Dimensions of Supply Chain Robustness: A Systematic Literature Review," *International Journal of Physical Distribution and Logistics Management*, Vol. 45, No. 1–2, 2015, pp. 118–137.

FederalSoup Staff, "House Measure Bans DOD A-76 Outsourcing Studies," *Federal Soup*, June 20, 2016. As of September 25, 2017:
https://federalsoup.com/articles/2016/06/20/dod-amendment.aspx

Fitzgerald, Bill, "Evaluating the Privacy and Security of Ed Tech," *The Journal*, April 14, 2016. As of October 25, 2018:
https://thejournal.com/articles/2016/04/14/evaluating-the-privacy-and-security-of-edtech-7-questions-to-guide-the-process.aspx

Freed, Rusty, and George Mollick, "Using Prior Learning Assessment in Adult Baccalaureate Degrees in Texas," *Journal of Case Studies in Accreditation and Assessment*, Vol. 1, 2010, pp. 1–14.

Gligor, David M., and Mary C. Holcomb, "Understanding the Role of Logistics Capabilities in Achieving Supply Chain Agility: A Systematic Literature Review," *Supply Chain Management: An International Journal*, Vol. 17, No. 4, June 2012, pp. 438–453.

Gligor, David M., Mary C. Holcomb, and Theodore P. Stank, "A Multidisciplinary Approach to Supply Chain Agility: Conceptualization and Scale Development," *Journal of Business Logistics*, Vol. 34, No. 2, June 2013, pp. 94–108.

Goldman, Charles A., Lindsay Butterfield, Diana Lavery, Trey Miller, Lindsay Daugherty, Trinidad Beleche, and Bing Han, *Using Workforce Information for Degree Program Planning in Texas*, Santa Monica, Calif.: RAND Corporation, RR-1011-CFAT, 2015. As of July 19, 2017:
https://www.rand.org/pubs/research_reports/RR1011.html

Gose, Ben, "How Colleges Cut Costs by Embracing Collaboration," *Chronicle of Higher Education*, March 26, 2017. As of October 25, 2018:
https://www.chronicle.com/article/How-Colleges-Cut-Costs-by/239580

Government Business Council, *Inside Federal Outsourcing: A Candid Survey of Federal Managers*, Arlington, Va.: Accenture, May 2015. As of October 25, 2018:
http://cdn.govexec.com/media/gbc/docs/gbc_government_outsourcing_report.pdf

Guidry, Leigh, "Higher Ed Increasingly Turning to Part-Time Faculty," *Town Talk*, July 10, 2015. As of October 25, 2018:
https://www.thetowntalk.com/story/news/local/2015/07/10/higher-ed-increasingly-turning-part-time-faculty/29954803

Harrington, Lisa M., Kathleen Reedy, John Ausink, Bart E. Bennett, Barbara A. Bicksler, Darrell Jones, and Daniel Ibarra, *Air Force Non-Rated Technical Training: Opportunities for Improving Pipeline Processes*, Santa Monica, Calif.: RAND Corporation, RR-2116-AF, 2017. As of October 25, 2018:
https://www.rand.org/pubs/research_reports/RR2116.html

Holliday, Marilyn, "AETC Announces Changes to Developmental Special Duty Assignments," Air Force Personnel Center, June 27, 2017. As of November 11, 2018:
https://www.afpc.af.mil/News/Article-Display/Article/1230618/aetc-announces-changes-to-developmental-special-duty-assignments/

Hurlburt, Steven, and Michael McGarrah, *The Shifting Academic Workforce: Where Are the Contingent Faculty*, New York: TIAA Institute, September 2016.

Hussain, Suhauna, "What Happens When a College Accepts Too Many Students?" *Chronicle of Higher Education*, August 1, 2017. As of October 25, 2018:
https://www.chronicle.com/article/What-Happens-When-a-College/240819

Karam, Rita, Charles A. Goldman, Daniel Basco, and Diana Gehlhaus Carew, *Managing the Expansion of Graduate Education in Texas*, Santa Monica, Calif.: RAND Corporation, RR-1899/1-CFAT, 2017. As of October 25, 2018:
https://www.rand.org/pubs/research_reports/RR1899z1.html

Kerekes, Carrie B., "Privatize It: Outsourcing and Privatization in Higher Education," in Joshua C. Hall, ed., *Doing More with Less: Making Colleges Work Better*, New York: Springer, 2010, pp. 235–247.

Klein-Collins, Rebecca, *Competency-Based Degree Programs in the U.S.: Postsecondary Credentials for Measurable Student Learning and Performance*, Chicago: Council for Adult and Experiential Learning, 2012.

Kleindorfer, Paul R., and Germaine H. Saad, "Managing Disruption Risks in Supply Chains," *Production and Operations Management*, Vol. 14, No. 1, March 2005, pp. 53–68.

Koedinger, Kenneth R., John R. Anderson, William H. Hadley, and Mary A. Mark, "Intelligent Tutoring Goes to School in the Big City," *International Journal of Artificial Intelligence in Education*, Vol. 8, 1997, pp. 30–43.

Lapovsky, Lucie, *The Higher Education Business Model: Innovation and Financial Sustainability*, New York: TIAA-CREF Institute, November 2013.

Lavastre, Olivier, Angappa Gunasekaran, and Alain Spalanzani, "Supply Chain Risk Management in French Companies," *Decision Support Systems*, Vol. 52, No. 4, March 2012, pp. 828–838.

Lewis, Matthew W., David M. Adamson, Jennifer L. Steele, Susannah Faxon-Mills, Mollie Rudnick, and Rick Eden, *Proficiency-Based Pathways: A Vision and Critical Elements for Achieving It*, Santa Monica, Calif.: RAND Corporation, 2013, not available to the general public.

Lumina Foundation and Gallup, *America's Call for Higher Education Redesign: The 2012 Lumina Foundation Study of the American Public's Opinion on Higher Education*, Indianapolis, Ind.: Lumina Foundation, February 5, 2013.

Magda, Andrew J., Russell Poulin, and David L. Clinefelter, *Recruiting, Orienting, and Supporting Online Adjunct Faculty: A Survey of Practices*, Louisville, Ky.: The Learning House, Inc., 2015.

Manacapilli, Thomas, Alexis Bailey, Christopher Beighley, Bart E. Bennett, and Aimee Bower, *Finding the Balance Between Schoolhouse and On-the-Job Training.*, Santa Monica, Calif.: RAND Corporation, MG-555-AF, 2007. As of June 25, 2019:
https://www.rand.org/pubs/monographs/MG555.html

Manacapilli, Thomas, Edward O'Connell, and Cheryl Benard, *Customized Learning: Potential Air Force Applications*, Santa Monica, Calif.: RAND Corporation, TR-880-AF, 2011. As of June 24, 2019:
https://www.rand.org/pubs/technical_reports/TR880.html

McFarland, Joel, Bill Hussar, Cristobal de Brey, Tom Snyder, Xiaolei Wang, Sidney Wilkinson-Flicker, Semhar Gebrekristos, Jijun Zhang, Amy Rathbun, Amy Barmer, Farrah Bullock Mann, Serena Hinz, Thomas Nachazel, Wyatt Smith, and Mark Ossolinski, *The Condition of Education 2017*, Washington, D.C.: National Center for Education Statistics, U.S.

Department of Education, NCES 2017-144, May 2017. As of September 29, 2017:
https://nces.ed.gov/pubsearch/pubsinfo.asp?pubid=2017144

McGrath, Cecile Hoareau, Joanna Hofman, Lubica Bajziková, Emma Harte, Anna Lasakova, Paulina Pankowska, S. Sasso, Julie Belanger, S. Florea, and J. Krivograd, *Governance and Adaptation to Innovative Modes of Higher Education Provision*, Santa Monica, Calif.: RAND Corporation, RR-1571-EC, 2016. As of September 29, 2017:
https://www.rand.org/pubs/research_reports/RR1571.html

Means, Barbara, Yukie Toyama, Robert Murphy, Marianne Bakia, and Karla Jones, *Evaluation of Evidence-Based Practices in Online Learning: A Meta-Analysis and Review of Online Learning Studies*, Washington, D.C.: U.S. Department of Education, September 2010.

Melnyk, Steven A., David J. Closs, Stanley E. Griffis, Christopher W. Zobel, and John R. Macdonald, "Understanding Supply Chain Resilience," *Supply Chain Management Review*, Vol. 18, No. 1, January 2014, pp. 34–41.

Miller, Trey, and Van L. Davis, *Leveraging Shared Savings to Promote High-Quality, Cost-Effective Higher Education*, Santa Monica, Calif.: RAND Corporation, PE-160-1-CFAT, 2015. As of September 29, 2018:
https://www.rand.org/pubs/perspectives/PE160-1.html

Molina, Arturo R., M. A. Velandia, and Nathalie A. Galeano, "Virtual Enterprise Brokerage: A Structure-Driven Strategy to Achieve Build to Order Supply Chains," *International Journal of Production Research*, Vol. 45, No. 17, September 2007, pp. 3853–3880.

National Center for Education Statistics, webpage, undated. As of September 29, 2017:
https://nces.ed.gov/

Person, Ann E., Jaime Thomas, Julie Bruch, Alexander Johann, and Nikhail Maestas, *Outcomes of Competency-Based Education in Community Colleges: Summative Findings from the Evaluation of a TAACCCT Grant*, Oakland, Calif.: Mathematica Policy Research, September 30, 2016.

Rendon, Rene G., "Outsourcing Base Operations Support Functions," *Program Manager*, Vol. 30, No. 1, 2001, pp. 16–31.

Sáenz, María Jesús, and Elena Revilla, "Creating More Resilient Supply Chains," *MIT Sloan Management Review*, Vol. 55, No. 4, June 2014, pp. 22–24.

Schmidtlein, Frank A., and Alton L. Taylor, "Identifying Costs of Instructional Technology in Higher Education," *Tertiary Education and Management*, Vol. 6, No. 4, December 2000, pp. 289–304.

Shapiro, Jeremy F., *Modeling the Supply Chain*, Belmont, Calif.: Brooks/Cole-Thomson Learning, 2001.

Sitzmann, Traci, Kurt Kraiger, David Stewart, and Robert Wisher, "The Comparative Effectiveness of Web-Based and Classroom Instruction: A Meta-Analysis," *Personnel Psychology*, Vol. 59, No. 3, Autumn 2006, pp. 623–664.

Smith, Albert B., Ronald D. Opp, Randy L. Armstrong, Gloria A. Stewart, and Randall J. Isaacson, "Community College Consortia: An Overview," *Community College Journal of Research and Practice*, Vol. 23, No. 4, 1999, pp. 371–385.

Steele, Jennifer L., Matthew W. Lewis, Lucrecia Santibanez, Susannah Faxon-Mills, Mollie Rudnick, Brian M. Stecher, and Laura S. Hamilton, *Competency-Based Education in Three Pilot Programs: Examining Implementation and Outcomes*, Santa Monica, Calif.: RAND Corporation, RR-732-BMGF, 2014. As of September 29, 2017:
https://www.rand.org/pubs/research_reports/RR732.html

Straus, Susan G., Michael G. Shanley, Maria C. Lytell, James C. Crowley, Sarah H. Bana, Megan Clifford, and Kristin J. Leuschner, *Enhancing Critical Thinking Skills for Army Leaders Using Blended-Learning Methods*, Santa Monica, Calif.: RAND Corporation, RR-172-A, 2013. As of June 24, 2019:
https://www.rand.org/pubs/research_reports/RR172.html

Sturgis, Chris, "Lesson Learned: Enabling Policy Isn't Enough, It Takes Incentives," CompetencyWorks.org, July 7, 2015. As of July 9, 2015:
http://www.competencyworks.org/policy/lesson-learned-enabling-policy-isnt-enough-it-takes-incentives

University of Washington, "Exploring the Pros and Cons of Online, Hybrid and Face to Face Education," *Leading Change in Public Higher Education: A Provost Report Series*, January 2013.

U.S. Code, Title 10, Section 688a, Retired Members: Temporary Authority to Order to Active Duty in High-Demand, Low-Density Assignments, December 12, 2017.

Williams, Richard, "AF 2013 Budget: Cues While Keeping Agile, Flexible, Ready Force," *U.S. Air Force News*, February 13, 2012. As of October 25, 2018:
http://www.af.mil/News/Article-Display/Article/111700/af-2013-budget-cuts-while-keeping-agile-flexible-ready-force/

Wisher, Robert A., "ADL: Foundations for Global E-Learning Interoperability," *Proceedings of eLearning and Software for Education Conference*, No. 1, 2011, pp. 33–41.

Xu, Di, and Shanna Smith Jaggars, *Examining the Effectiveness of Online Learning Within a Community College System: An Instrumental Variable Approach*, New York: Columbia University, CCRC Working Paper No. 56, April 2013.

Yarnall, Louise, Barbara Means, and Tallie Wetzel, *Lessons Learned from Early Implementations of Adaptive Courseware*, Menlo Park, Calif.: SRI Education, April 2016.

Lightning Source UK Ltd.
Milton Keynes UK
UKHW051843310820
369132UK00007B/52